AuthorHouse™
1663 Liberty Drive
Bloomington, IN 47403
www.authorhouse.com
Phone: 1 (800) 839-8640

Published by AuthorHouse 07/15/2016

ISBN: 978-1-5246-1848-3 (sc)
978-1-5246-1849-0 (e)

Library of Congress Control Number: 2016911494

Print information available on the last page.

authorHOUSE®

The First and Original Inventor

Human Powered Transportation Means for the 21st Century

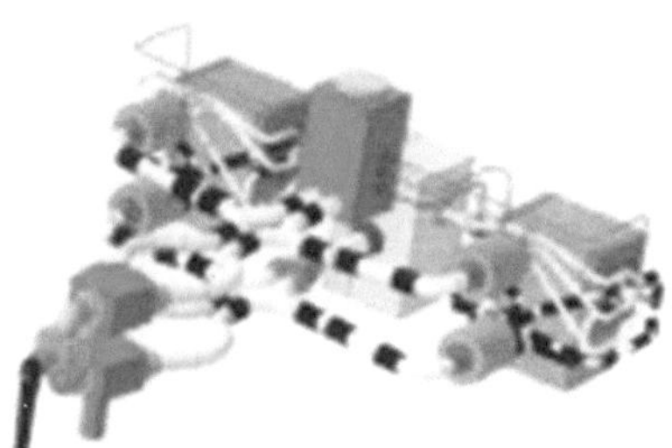

Self-propelled Motor

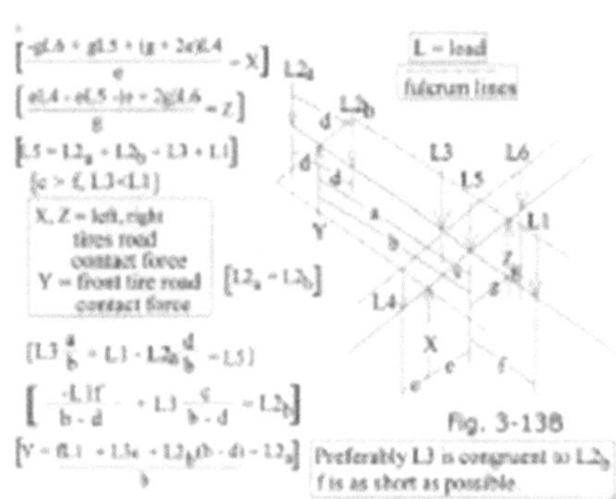

Landing Gear Equations

By Richard Chastain M.E.T.

Volume 2

A Cut Above Mechanical Engineering Technology and Short Stories

Dear reader: So far you have made it to Volume 2 of this two volume publication, congratulations. This volume, however, does not include the evolution of these ideas and discoveries beyond their rationale of analytical concept in mathematics and design drafting using CAD software. Funding of this or these project(s) is estimated to cost at least as much to have a machine shop make the parts for the least expensive of the concepts as it would cost to purchase all the machine tools necessary to make the concept myself. I hope you find the remainder of this publication interesting and enjoyable. Thanks for reading it, or just looking at the pictures, and for studying it if you choose to do so.

Index

The Seven Years Year of Relaxation

It is 2070A.D. and the cost of living has risen to a point where a gallon of milk costs $12. The population is 7.8 billion on the Earth. The population of the United States alone is 1.02 billion people. Decades ago an individual brought an idea to the Government that was centuries old. The Government ignored it. As an individual researched old documents sent to the Government an old document was come across, and brought to the Congress and to the State Senator from which State the document had come those many decades ago. In the document was written a suggestion of how to gain from the loss of control over the economy and how to put it in to some kind of order. The problem facing the United States under the suggestion was that it would have to submit to "getting small". These are the conditions of the suggestion; it is called the Seven Years Year Of Relaxation:

In the present day there are many loans out which are subject to many years, and these loans are spread out without regard to their start date or their end date, except that they begin and end, and they are stretched out over numerous periods of time, each one different, starting on a different day it may be for the same time, starting on the same day for a different time period, and these loans are smeared over the calendar as it were a gray mass, a smudge of loans, for years and years, beginning and ending at all times, every day, ceaselessly, and they are routinely started up without regard for the reduction of paperwork, or for making any concession to account for the smearing of loans over time otherwise.

There is a solution: Beginning immediately, set a seven year period throughout which any loan may be taken out. The following day the period for which a loan may be taken out is one day less than seven years. Then, the following day, two days less than seven years, then three, four, five less, and so on, and the days count down as the seven years elapses. Borrowing the absolute maximum amount of money one can borrow at the beginning of the seven year period does not allow the same amount the next day unless the interest rates go down. This is ad Perpetua continuously as the days progress towards the end of the seven year period. Anyone can borrow money, but if anyone borrows as much as one can borrow, the next day if the interest rates are the same that one cannot borrow as much as the day before because there are less days to pay it back before the seven year period ends.

All present outstanding loans will have to filter out provided the seven year period is established. The thirty and forty year loans will eventually fall out after they expire. Once they are all gone then no one will be able to take out a loan longer than seven years provided it occurs at the beginning of the seven year period; otherwise the period will be less long as the seven years period counts down from seven years to zero, at which point, if all the loans are paid back there can be a celebration overnight or on the weekend if the seven year period ends on a Friday and the banks close for the weekend.

Nevertheless, if the richest person in America were to apply for a loan, they would not be able to borrow as much per each day the seven year period counts down to zero days each day the seven years period becomes less and less in days. No matter how rich that person is, unless the interest rates go down they can only borrow less than the day before or else they cannot afford the loan. That is where "getting small" plays in, downsizing the United States. That's where the money is.

So now there is the prospect of someone not being able to pay back a loan. Prior to the last and final year of the seven years year of relaxation it should seem logical to foreclose on the loan: however, if the individual has paid the loan for the past six years and has taken out a seven year loan, and in the final few months of the seven year period cannot pay the loan payment, the loan shall be forgiven of the person not able to pay the loan who has shown such willingness to pay the loan for so long and the loan shall be forgiven the person, and the end of the seven years year of relaxation shall not be a miserly disgruntling situation of failure to repay a debt but shall be a celebration of all debts paid and for what it's worth Life, and the debt shall be forgiven. As well, in the final year it shall be allowed them that they shall pay as best they can and if in the end on the last day they cannot pay back the debt it shall be forgiven them.

Now most people will want good credit in the next seven years period and a forgiven debt shall not go down as bad credit, because it was in the last year of the seven year period. Others however, who do not repay their creditors may get bad credit reports for not paying back their loans in the prior seven years save the last one of the seven years period. I will leave these persons to the jurisprudence of the courts and with their consideration that money is not that important although one needs it to eat and cloth oneself, besides that keeping-up-with-the-Joneses isn't a reason to default on a loan, it is reasonable to assume that Judges can decide these things, therefore I will continue:

The prospect of having a seven years year of relaxation introduces interest rate competition of lending institutions as the seven year period comes to a close, in order to have people take out loans from them. Interest rate competition is good for the economy provided the interest rates go down, I assume. Of course the interest rates could always go haywire at the end of the seven year period. Who knows? Getting everybody to "Fall-in" as it were, and get all the outstanding loans paid while the seven years period is going on and having them all fall-in to the seven year period and making everyone and everything "Get Small" (Downsize) will be an accomplishment in itself. Rich people will have to live in smaller houses because they can't borrow as much money as they could yesterday during the seven years period because they have less number of days to pay off the loan. Paperwork will be consolidated in to seven year periods with no paperwork at the end of the seven years, except maybe all the people who prepare for the seven years loan and get approved to be started on the first day of the seven year period. Maybe there will be lots of paperwork: nevertheless, there will be a celebration at the end of the seven years, whether it be overnight or over the weekend every seven years, a celebration of no loans outstanding, no one owing any money. The thing is that not being able to borrow as much per consecutive days leads to wealthy people having to buy small, and smaller every day that passes. Unless there is more money to borrow with there can be no borrowing more money as long as the period of time is decreasing to borrow money in, if the interest rates are not competing and going down as the days are becoming fewer and fewer for borrowing the money as the seven years period elapses.

Yet the Government will not heed this interest, at least not yet anyway. It may yet be one day when the Government sees that it is feasible to stop being so intimidating when it comes to the malfeasance of money and proliferate it down through to the people that money is not all that important with the exception of causing someone's death by taking it. As for loans, Ebenezer Scrooge is the example in this respect. So Bah Humbug to money. The seven years year of relaxation should make it so there is plenty of money for everyone and there eventually should be no more poor people in the United States. Take it upon oneself to loan someone $50 under the seven years year of relaxation. Count down from 2556 days starting your own seven year period in the event that anyone should want to borrow money from you. Now set your interest rate to, say, x 1.5 after seven years. That's almost $1.82 a month at fixed interest. Seven years is not a long time to pay back a loan except to some people. Now this is where getting borrowing as much as possible is not as much as what borrowing is today. Say, the richest person in town could borrow $30,000,000 today at 5% interest and pay it back in 30 years. Under the seven years year of relaxation that person might only be able to borrow $250,000 at 5% and pay it back in seven years, or $249,300 the next day at 5% and pay it back in 2555 days, or some such thing like that. As the days decrease in number the person cannot borrow as much as the day before at the same interest rate. The payments will either be more for the same amount or the person has to borrow less to get the same payment because they have less time to pay off the loan. The seven year period counts down from seven years.

Let us celebrate when the seven year period is over and there are no outstanding loans. What a day that will be. And gradually all the people of the nation will get wealthy.

Prices should also be competitive towards the end of the seven year period and borrowing large amounts of money becomes less, prices would seem to suggest a going down, which perhaps would seem to be a notoriously bad effect on the economy, but that would seem to introduce more spending which would be good for the economy. All this competition would with interest rates and prices would seem good for the economy. Capturing the loans period in to perpetuating seven years periods running consecutively indefinitely should capture and get gain from putting a start date and an end date to the period in which loans may be made. Shorter loans may be made during the seven years period and paid off provided they start and end during the remaining time during the seven years period or the remaining days of the seven years period: e.g. a four year loan may be made within the period remaining of five years and 234 days in the seven years period.

Not all explanations are made here for the Seven Years Period. Competition is a good thing, so is downsizing. It does seem to be proven that some businesses prosper and manage to stay in business by downsizing. The National Debt seems to be a problem. No President, it seems, seems to be able to get a good handle on the National Budget. It would seem logical to establish a test group for the Seven Years Year Of Relaxation in the United States and try it and see how it performs, and if it performs well enough expand it in the United States, and should it prove normal as a performing factor in economics let it take over the present status of the lending institutions in the United States if things are working out with it and let's "get small" (downsize). Perhaps downsizing will introduce recovery in the National Budget. Let us hope so.

I hope you've enjoyed this Thesis despite the fact that it was a dollar and only eight pages long. There is probably somewhat more that can be written about the Seven Years Year Of Relaxation but I haven't thought of what it could be just now. If I think of anything else I will write it and reapply it to the file download. Thanks.

This Thesis was written by Richard L. Chastain

Human Powered Helicopter Explanation

Richard Chastain (author)

Human Powered Helicopter Explanation

(VTOL Pedaled Flying Machine Powered By A Hydraulic Power Plant)

This human powered helicopter is designed to lift a 275+ pound person. The calculations for the lift blade apply 2,431.6 pounds of lift and drag to the lift blade at maximum output. Lift blade revolutions output are shown in the calculations. A calculation of torque applies force to the impeller inside the lift blade impeller gallery as shown in the calculations. The formula for Fluid Mechanics of Inertia applies the required volume of hydraulic fluid to account for the static load of force at the impeller times revolutions plus fluid flow through the relief passageways at static flow volume output at maximum applied force. This sum quantity equals the initial meshing gear teeth volume of the secondary power transmission. The same force at the lift impeller applies at the secondary power transmission. Note that all of the hydraulic forces are in SUCTION. Next, the clearance volume in the secondary power transmission is calculated at the same force, the exponent of the formula for Fluid Mechanics of Inertia is multiplied to the clearance volume and the product is added to the initial meshing gear teeth volume of the secondary power transmission. This is the final volume of the secondary power transmission. *Note that all these calculations are made with respect to two crank revolutions at the operator pedals and chainring per second which is equal to maximum output performance, 128 cadence per minute.*

The applied force is still the same at the secondary power transmission meshing gears teeth. Now, roughly estimate a circumference for the secondary power transmission impeller and calculate its torque using the force at the meshing gear teeth of the secondary power transmission center gear diameter (preferably the working depth diameter). The force at the impeller radius working depth of the impeller blade vanes is now calculated back through the drive train to the operator. Once the desired Operator force is applied divide a percentage of the desired force at the operator (say, 67%) into the resulting force of the calculations at the operator and multiply the quotient to the diameter of the impeller and the product should result in the desired diameter of the impeller resulting in the 67% required operator's force through the drive train at the pedals center axis. The 67% applied force will equal 100 minus the tail rotor force divided by the lift blade force times 100 equals the desired percentage of primary power transmission operator's force.

The applied force at the secondary power transmission is now calculated in torque to be the desired force at the resulting diameter of the secondary power transmission impeller diameter. Use the formula for Fluid Mechanics of Inertia to calculate the fluid volumes of static load volume of the impeller blade vanes times revolutions plus the relief passageways volume at static load exponent, and this will be equal to the primary power transmission initial meshing gear teeth volume. The applied force at the secondary power transmission impeller conveys to the primary power transmission and is the same force. Now calculate the clearance volume in the primary power transmission and using the force at the secondary power transmission impeller blade vanes and the formula for Fluid Mechanics of Inertia calculate the static volume of the clearance volume of the primary power transmission at the applied force using the exponent of the formula for Fluid Mechanics of Inertia and add the product to the initial meshing gear teeth volume of the primary power transmission initial meshing gear teeth volume. This will be the final meshing gear teeth volume of the primary power transmission.

The Tail Rotor

The applied suction at the lift impeller will make the helicopter want to rotate the airframe clockwise looking down from the top. Again, torque applies at the radius of the lift impeller with respect to the axis of the tail rotor. Keep in mind that all forces are applied with zero motion although motion is accounted for as revolutions in calculations. The same formulas apply with respect to the tail rotor blade vanes and with the lift rotor blade vanes although the tail rotor blade vanes will pitch to make changes for yaw control, increasing and decreasing or reversing the amount of lift plus drag force created by the tail rotor blade vanes. The entire force at the tail rotor (lift + drag) is used in calculations to determine the dimensions of the impeller and the power transmission meshing gear teeth volume of the tail rotor power transmission assembly exclusively.

The impeller for the tail rotor is calculated to turn the required output revolutions by minimizing the frontal area profile of the tail rotor impeller blade vanes while managing the secondary power transmission of the tail rotor impeller so the impeller of the tail rotor secondary power transmission and primary tail rotor power transmission are all kept limited to minimum tolerances. Varying the diagonal frontal area profile of the impeller blade vanes and the impeller's diameter will align the tail rotor secondary power transmission to minimum tolerances. The applied force at the tail rotor impeller (lift plus drag (lateral and transverse)) is calculated back through the drive train to the operator and added to the initial operator's force (for example: 33% tail rotor force + 67% lift blade force = operator's force). The sum of the two forces at the operator should be the desired operator's force. The diameters of the secondary power transmission impeller and tail rotor secondary power transmission impeller must be aligned to achieve the desired operator's force.

The sum of the two forces is applied at the operator pedals and must be within the operator's tolerances. Hydraulic fluid should be sewing machine oil or some likeness of thin lubricating oil of low viscosity, perhaps linseed oil but must not break down over time and not be hygroscopic. Most parts are to be made of elastic plastic or some other durable light material that is inflexible. The control linkage is a simple gear driven bell crank and connecting rod assembly.

The tail rotor spins counterclockwise looking at the tail rotor from the left side of the helicopter. The tail rotor mechanism consists of a ball jointed bell crank which turns a screw that operates a lever with a forked end that moves a slider (the internal component of which spins on the tail rotor output axle) that operates the blade vanes to pitch, whereby the lift and drag of the tail rotor are controlled and so yaw occurs. The tail rotor blades turn on an axle that is engaged with the tail rotor, and output impeller. The screw threads of the screw that operates the lever are left hand threads for the required blade pitch with the steering and control linkage.

The Secondary Hydraulic System

All of the hydraulics have reservoirs. The secondary hydraulic system operates by syphoning hydraulic fluid from the reservoirs. The purpose of the secondary hydraulic system is to prevent air from leaking in to the entire assembly at the axes of rotation through the grease seals. Since there is suction throughout the entire assembly during operation under load there is continuous suction at the axes of rotation at the grease seals and hydraulic fluid will be sucked in from the reservoirs through the secondary hydraulic tubing instead of there being air sucked in at the grease seals. By applying secondary hydraulic tubing to suck hydraulic fluid under suction force the possibility of sucking air in at the grease seals is eliminated. Hydraulic fluid is continuously replenished in the reservoirs by the continuous recirculation of hydraulic fluid in the assembly.

There are six reservoirs on this helicopter.

Pitch and roll are controlled by the operator leaning forward and backward and from side to side. There are four locations for counterweights at the rear of the helicopter. The counterweights are bullet shaped with a threaded end and hang down by their threaded end. The counterweights will probably be to offset the weight of the hydraulic fluid. A counterweight to offset the weight of the pilot has not been added at this time. No account has been taken for lift blade vanes pitch control at this time.

There is a throttle. The throttle is hand operated by a twist grip mechanism at the right hand on the handlebars. The throttle controls the flow of hydraulic fluid to the lift impeller by closing off the bypassing fluid flow bypassing the lift impeller gallery by channeling it through the throttle and allowing more hydraulic fluid to flow to the lift blade impeller gallery thereby applying more hydraulic fluid flow to the lift blade impeller gallery increasing the force of flow to the lift blade. The operator may crank the pedals and vary the cadence while operating the throttle and control the lift blade output revolutions and the cadence by manually operating the throttle, to be able to ascend and descend in a more controlled manner and at a more comfortable cadence.

The control of the lift blade vanes pitch is now being engineered. This consists of articulated lift blade vanes having ball joints with connecting rods which control the pitch joining to ball joints mounted in a swash plate which swash plate is articulated around a ball joint mounted to the main shaft to the lift blade. The swash plate is manipulated by control rods with respect to the forward and aft (pitch) blade vanes pitch control by a bell crank, and the left and right (roll) blade vanes pitch by independent bell cranks. The rotating swash plate is kept from twisting caused by the drag force of the blade vanes during lift, by fingers in the rotating swash plate that extend in to the central ball joint on four sides and by pins in the ball joint that protrude in to the main shaft ring made for them. There is only about 100 pounds of drag force being applied to the rotating swash plate through the connecting rods joining the lift blade vanes to the swash plate. The entire mechanism is operated by control rods and bell cranks to the operator's control stick which is an independent bell crank control lever in two axes with a head tube at the two levers axis which axis will swivel about on a universal joint where the two axes of the levers' rotation intersect. The steering yaw control rotates through the universal joint to the yaw control mechanism to change the tail rotor blade vanes pitch (yaw). The operator moves the handlebars in a circular manner to control the lift blade vanes roll and pitch pitch control and the arc levers at the base of the handlebars stem (at right angles to one another) operate the bell cranks that control the swash plate.

This concludes the explanation of the human powered helicopter.

The Human Powered Helicopter

The Human Powered Helicopter

By Richard L. Chastain

The control of the lift blade vanes pitch is now being engineered. This consists of articulated lift blade vanes having ball joints with connecting rods which connect the articulation joining to ball joints mounted in a swash plate which swash plate is articulated around a ball joint in the center mounted to the main shaft to the lift blade. The swash plate is manipulated by control rods with respect to the forward and aft (pitch) blade vanes articulation control by synchronized bell cranks, and the left and right (roll) blade vanes articulation by independent bell cranks. The rotating swash plate is kept from twisting caused by the drag force of the blade vanes during lift, by fingers in the rotating swash plate that extend in to the central ball joint on four sides and by pins in the ball joint that protrude in to the main shaft ring made for them. There is only about 100 pounds of drag force being applied to the rotating swash plate through the connecting rods joining the lift blade vanes to the swash plate. The entire mechanism is operated by control rods and bell cranks to the operator's control stick which is an independent bell crank control lever on two axes with a universal joint at the two levers axes intersection which center will swivel about on a universal joint where the two axes of the levers' rotation intersect. The steering yaw control rotates through the universal joint to the yaw control mechanism to change the tail rotor blade vanes articulation (yaw). The operator moves the handlebars in a circular manner to control the lift blade vanes roll and pitch blade vanes articulation control and the arc levers at the base of the handlebars stem (at right angles to one another) operate the bell cranks that control the swash plate. The operator rotates the handlebars clockwise and counterclockwise to control the tail rotor blade vanes articulation.

The tail rotor blade vanes articulation is controlled in the same way as the previous design of human powered helicopter. The mechanism has not been changed much, only the design appears to be similar. The only drawback remaining in the entire assembly is the flexing of the connecting rods during operation since there now is only a single connecting rod where there were once pairs.

The Power transmission units and the impeller galleries are to be made of polyurethane surrounded by epoxy-carbon fiber with metal inserts for bearing races and containing the gears in the power transmissions. Any other component which may need a metal piece may be applied. Bolts will be inserted with threaded inserts in to the polyurethane and epoxy-carbon fiber for mounting and assembly. The airframe will include bearings to support the impeller gallery output shafts to prevent deformation of the epoxy-carbon fiber and polyurethane constructions, as well the secondary power transmission units will have bearings supported by the airframe. The primary power transmission units' chainrings will have bearings exterior to them to prevent deformation during loading. These bearings are placed at the chainrings where the applied forces are critical. The calculated rate of the impellers for both the primary impeller and the tail rotor primary impeller are four revolutions per crank circuit of revolution. In the previous design of human powered helicopter the rate of impellers revolutions was sixteen revolutions per crank circuit of revolution. Also in the drawings and in the calculations the lift impeller blade vanes profile area is calculated at ¼" square and not ½" square to calculate the secondary power transmission unit meshing gears teeth depth.

The center gears are to be made of titanium, not to say that the center gears need to be tough but that the output shafts need to be strong and are directly tied in to the gears. The center gears and output shaft are made as one piece. The planet gears may be made of aluminum plated with stainless steel. The output shafts for the lift blade and tail rotor are also titanium. The blade vanes hubs for both are titanium. The lift blade spars are titanium. The lift impeller is titanium. The tail rotor impeller can probably be aluminum but if it is titanium it will weigh less although it will cost more. The airframe is 6061-T6 aircraft aluminum but may have to be made of reinforced fiberglass to reduce the weight of the helicopter while hopefully not sacrificing too much strength.

To date, the chain of the chain drive is motorcycle chain, ½ inch width and ¾ inch pitch, which may be stainless steel. The 72 tooth sprocket is a freewheeling sprocket freewheeling clockwise when viewed from the left side. This will allow the operator to crank the pedals crank while the secondary impeller begins to turn. By design calculations the fluid flow to the secondary impeller is greater than is required to turn four revolutions per crank due to the clearance flow volume and the secondary impeller should transfer that extra revolutions force to the 72 tooth sprocket to assist the operator in cranking. However, this may turn in to a runaway mechanism and a throttle will be need to be returned to the assembly. A relief bypass flow may need to be employed to control the runaway problem, if there is one. Sprocket spacing at the primary power transmission units is 10 ½ inches in two dimensions. Sprocket spacing between the 72 tooth sprocket and the primary impeller is approximately 165 inches to allow for chain slack so the chain doesn't break. Sprocket spacing between the 52 tooth chainring at the operator's crank and the flywheel also allows for slack so the chain does not break: however, chain slack in the flywheel drive should be as little as possible to sustain the maximum performance of the flywheel at the operator's crank.

In the drawings there are 36 tooth sprockets at the all the power transmission units. This calculation has been in error and their ratio to the center gear in the power transmission units is 4.5 : 1. With 24 tooth sprockets at the power transmission units the ratio at the center gears will be 3 : 1 which is what the original calculation were supposed to be at. Also the 25 tooth sprocket at the crank will be changed to a 24 tooth sprocket, leaving the 36 tooth sprocket at the crank. All these sprocket changes will affect the spacing between the power transmission units so that the chain will align on the sprockets. I haven't calculated the spacing between the sprockets yet and will not redraw this first drawing of the human powered helicopter but I am going to draw a new version of the human powered helicopter.

Getting the weight down of the helicopter is a primary consideration.

The human powered helicopter is being redesigned. The power transmissions are being located directly under the lift blade; the collectors from the primary power transmissions are being located directly forward of the power transmissions forward of the lift blade axis or rotation. All other components are normal as previously drawn. The aircraft has been shortened by 13 feet seven inches by locating the power transmissions vertically under the lift blade axis. Also, the lift blade primary power transmission units are on the bottom of the stack, the tail rotor primary power transmission units are lighter and are therefore on top of the stack of power transmission units located vertically under the lift blade axis. The primary impellers are both the same from the previous model's design. Their calculations have been adapted. Only the primary power transmission of the tail rotor has been changed with the secondary tail rotor power transmission because the tail rotor is moved from 40 feet out to 27 feet out so the calculations had to be redone and the power transmissions had to be recalculated. The primary impeller for the tail rotor has remained the same from the previous model's design.

Hopefully still, the human powered helicopter will not turn out to be an elaborate ceiling fan. The success of this latest design will be on the lightness of parts. The appearance of the assembly design is representative of the parts' locations. The aircraft must be constructed to meet FAA certification for airworthiness and certificate of airmanship. The end result may not even appear to be the same constructed materials as the drawing portrays them. The airframe may have to be made of plastic because if it is aluminum it will most probably turn out to be too heavy. The output shafts however will have to be Titanium because of their lightness and strength in the places where they are applied, as will the impellers, because of the forces. Torque is very high and forces are great in the places where the Titanium is applied. A minimum of Titanium is ascertained in the design to keep down costs and applied where necessary while aluminum is applied otherwise. The assembly drawings of the human powered helicopter is just a rendering, the actual end result of the parts designs will be different once the values of applied forces in static equilibrium are evaluated for limits of tolerances on parts geometries for shear areas at absolute minimum elastic limit at applied maximum force and the aircraft is constructed to its minimum possible weight. This capacity however may take 100 years to file out of the original prototype (hone to exacting specifications), so I am not expecting miracles out of the original prototype, but it can happen if the right people and today's technology is applied.

This latest version of human powered helicopter has an error in it. The tail rotor impeller is too small resulting in applying too much force to the operator at the crank. As you can see I used previously designed parts to apply to the latest version of the human powered helicopter and the tail has been shortened by several feet increasing the force at the tail rotor without having changed the impeller diameter. Correct this error. The applied force at the operator is about 250 pounds at take off with it as it is.

Hopefully the human powered helicopter once it is finally designed will be light enough to include an electrical system including an alternator, a 12 volt battery, all the accoutrements necessary for driving the 12 volt electrical system and charge the battery, a voltage meter/charging circuit, and amp-meter, an altimeter, two transceiver multi-channel radios with output power/range control, running lights, landing light, strobes, oscillating red light on the tail, an interior light, and an air speed indicator, and circuit breakers. Not much else is really necessary for a human powered helicopter except maybe hydraulic fluid levels in the reservoirs and vacuum sensors in the hydraulic lines, with indicator meters. Warning lights for all the essential points would be considered convenient, and alarms would be convenient as well, as this helicopter is manually controlled there will probably be no need for a computer to monitor anything. Warning indicators should be sufficient.

I haven't finished this last attempt at the human powered helicopter either. It needs control linkage to the lift blades from the operator, the chain drive linkage, the remainder of the tubing which can be drawn using the other human powered helicopters I have drawn and the self-propeller motor tubing sizes, and it needs handlebars, some other things, and a throttle. This throttle however may be linked between the suction of the primary impeller tubing and its head outlet tubing instead of between the inlet and outlet tubing of the lift blade impeller. Thanks. I will finish drawing this latest attempt to draw the lightest human powered helicopter of the three eventually.

Perpetual Motion Engine Explanation

Richard Chastain (author)

Perpetual Motion Engine Explanation

Assembling the power transmission

To calculate the size of the power transmission meshing gear teeth volume the size of the impeller has to be described by choice. The application of fluid flow volume to the impeller can either be by a multiplier, by a multiple of two, or by a force in pounds, preferably in a multiple of two simplifying the analytical procedure. In this case 24 is the multiplier used to calculate the gears depth. Without using the formula for Fluid Mechanics of Inertia and the density of the hydraulic fluid, calculate the volume of the impeller gallery less the impeller volume, impeller bracket gasket and impeller bracket and multiply by 24. Calculate the impeller blade vanes volume and multiply by 24 x 1.5. Divide the sum of the total volumes by the area of the meshing gear teeth of the power transmission, using one area and multiply that one area by the number of meshing gears. This is the final meshing gear teeth volume for a 24 multiplier power transmission. Then divide by two power transmissions. Fabricate gears (center and planet gears are equal depth) according to these calculations. Fabricate the power transmission casing to the same depth. A +0.007" gasket separates the power transmission casing from the power transmission plate. The power transmission plate contains the ball bearings for the center gear and the pilot bearings for the planet gears and the openings for the fluid flow tubing, and bolt holes for the power transmission grease seal. Assemble the power transmission. The grease seals fit over the center gear output axle and bolt to the power transmission plates with a gasket between them. Assemble the power transmission to the mount.

Assembling the impeller assembly

In this case apply a 3 inch radius to the impeller. This is the working depth of the impeller blade vanes. These impeller blade vanes are ¼ inch by ¼ inch in the profile view. The total radius of the impeller is 3 1/2 inches at the impeller blade vanes. The impeller is ½ inch thick at the impeller blade vanes but the blade vanes are only ¼ inch thick leaving 1/8 inch to either side of the impeller blade vanes for clearance volume. Apply a 3 1/2 + 1/16 inches radius to the outside of the impeller and extrude to depth to one side approximately 1 inch deep, this leaves 1/16 inch clearance above the impeller blade vanes. To the other side apply the same radius to an extruded depth of ¾ inch. The depth of the deeper side of the impeller will have to be adjusted as the impeller casing is fabricated. Fabricate the casing of the impeller to include 0.007" radius clearance (+0.014" diameter) for the impeller to fit into the casing. The grease seal and gasket fit over the center gear input axle, then the impeller casing fits over the input axle of the center gear and the bearing goes in and bracket fits over the end of the center gear output axle and is attached with an E-clip. The impeller brackets attach to the center of the impeller on one or the other sides with a +0.007" rigid gasket between the bracket and the impeller to keep hydraulic fluid from leaking out through the bracket at the center gear input axle. The bearing rides on the bracket. Clearance is machined in to the impeller casing for the bracket, gasket and bolts, and a space is machined for a bearing in to the impeller casing. The remaining bore through is +0.014 " the diameter of the bracket diameter that will stick through the impeller casing. The depth of the casing for the impeller will allow the fluid flow ports to be drilled out to allow for heavy diameter and heavy wall thickness tubing, this is in the event that if there is a given load to the perpetual motion motor and the suction force increases the tubing will not crack. The holes drilled in the casing must have enough material to one side (the open side for the impeller to enter) so the drilling operation will not deform the material at its thinnest area to the open side. Therefore, the impeller depth to one side must be aligned to the plane of the impeller casing depth while the impeller blade vanes and clearance area and fluid flow ports centerlines must align keeping the +0.007 inch clearance all around.

The impeller grease seal and gasket go on the other center gear input axle, then the impeller plate fits over the same center gear input axle and the bearing goes in, and the other impeller bracket goes on the end of the center gear input axle and attaches with an E-clip. A 0.007" rigid gasket goes between the bracket and the impeller and the brackets are then bolted to the impeller. A +0.007" gasket goes between the impeller casing and the impeller plate and the two halves of the impeller assembly are fitted together. The same bores are taken for the bracket to the impeller on the plate side including the bearing. The entire assembly is mounted on the mount as construction is proceeding.

Assembling the tubing and perpetual motion

The elbow tubing is placed first at its respective angles according to the drawings. The straight tubing and elbow tubing are joined by rubber sleeves with hose clamps and the tubing are pressed as close together as possible to prevent an aneurism from occurring in the rubber sleeve at the gap in the tubing where the tubing comes together. The tubing is butted together. The tubing fits in to the collectors with rubber sleeves. The tubing is glued or soldered in to the impeller plate fluid flow ports. The throttle also is fitted with rubber sleeves. The secondary hydraulic tubing is fitted from the grease seals to the reservoir and fixed in place. Hydraulic fluid is drawn in to the grease seals from the reservoir through the secondary tubing as suction is drawn by operation of the impeller and with load as operation applies suction. Hydraulic fluid is recirculated back in to the reservoir. The throttle tubing is assembled the same way. The secondary hydraulic system does reduce the probability of power slightly but it is the only way to seal in the hydraulic fluid. The reservoir is filled, a crank is inserted in to the outboard end of the center gear (on the right side) and the crank is turned clockwise as the throttle is closed, hydraulic fluid flow is channeled in to the impeller gallery by suction and away from the throttle (throttle closing), the amount of force of suction being applied at the meshing gear teeth is the same at the impeller suction fluid flow port; there is a 3:.95 inch radius ratio of torque where the radii of the two forces are equal (between the impeller blade vanes working depth and the meshing gear teeth circumference). As the throttle is closed the flow volume to the impeller increases increasing the torque while the forces remain equal. The force of torque being applied would seem that the sum of the moments greater than zero is achieved. This device can be varied by applying the gears depth with respect to the radius of the impeller (changing the multiplier), the impeller clearance volume around the impeller blade vanes, and the blade vanes volume. Some other variables may apply like gear teeth count for smoothness and minimizing friction, but the author is unable to constitute a reasonable faction for why this perpetual motion motor will not work except that perpetual motion motors are notorious for not working.

Update

Lately the tubing diameter has changed. The areas of the tubing have been equalized so the sum of the areas of the six inlet ports of the collector are equal to the area of the outlet port of the collector. The related tubing area have all been made the same diameter. The only change in area occurs at the port area of the impeller gallery at which point the fluid flow is required to speed up anyway. This tubing size change relaxes the hydraulic load on the fluid flow. Still, hydraulic fluid flows to the inlet and outlet ports of the impeller gallery only with less haste and there is virtually no pressure or suction load compared to what is inside the impeller gallery. Pascal's Law of areas in hydraulics makes an equal amount of volume displace from six in to one out with equal area for both. I suppose the displacement distance ratio is 1:1 if the areas are equal.

Here is a problem: Suppose head and draw forces are equal at the meshing gear teeth at 7 + 7 = 14 at the radius of .95. The radius of the impeller is 3 and has a force of only the one draw force of 7. $3 \times 7 / .95 > 14 \times .95 / 3$. Even if there is no head force to the impeller there is rotation in the torque lever arm. All the head force is expelled in the reservoir. Even more probable is there is very little head force at the meshing gears teeth head outflow. Only the viscosity of the hydraulic fluid will introduce head force. Figure out some way to prove that this motor applies with respect to the sum of the moments equal to zero and I will admit that it doesn't work. No friction is included in this equation but will be included in the actual working model.

For reference, all the previous perpetual motion motors I have seen in my life growing up spilled all their hydraulic fluid out in their attempt to perform; my perpetual motion motor's hydraulic fluid is all contained and none of it spills out. Also, this is not physics this is statics.

It seems impossible but even when the moving parts are held in static equilibrium and are kept from moving and the fluid forces are applied at their respective areas the torque equation still applies the sum of the moments greater than zero with the throttle fully closed. It may be required that the throttle be nearly fully closed for the machine to operate. With the moving parts released from static equilibrium and allowed to move freely and the throttle is closing mostly all of the way, and the parts are applying their respective forces all simultaneously, what could be causing the operation to fail? Discover the solution to this dilemma and resolve it by conforming the device to accommodate a solution to its operating principles. There must be some cause for why it won't work, the cause just hasn't been discovered yet. The solution for why it won't work has not been found. Only the areas ratio of the impeller gallery inlet and outlet ports and the meshing gear teeth inflow and outflow ports is left to question.

Even on the atomic level an atom or molecule of hydraulic fluid applying a force in the inflow and outflow of the meshing gears teeth (twelve atoms or molecules respectively (twelve inflow and outflow ports individually at the meshing gears teeth)) will apply six molecules or atoms at the inlet and outlet ports of the impeller gallery at the same force that is being applied at the meshing gears teeth and the sum of the moments will be greater than zero. Remember that the equation being applied to the impeller blade vanes is the equation for fluid mechanics of inertia: hydraulic fluid density x blade vanes displacement volume x 2 raised to the power of the quantity of the applied force divided by the quantity of the hydraulic fluid density x the blade vanes volume close quantity, minus 2 to the nth power, then divide by 2 to the same nth power, then close the exponent quantity and subtract the exponent quantity by the same n value, equals the hydraulic fluid density x the blade vanes volume x 2 raised to the power of p, which product is equal to the applied force in pounds. The applied force is the same force that acts in the meshing gears teeth, and if it is double for inflow and outflow for it to be the same force at the impeller gallery inlet and outlet ports simultaneously the equation for the sum of the moments is still greater than zero. The calculation for the meshing gear teeth volume is taken from the calculations from the impeller gallery volume to the meshing gears teeth volume as written at the beginning. This is complicated and the subject is beginning to go around in circles, no pun intended. Good luck. It is proving very difficult to make this perpetual motion motor not work.

And, once again, the author is financially incapacitated and cannot function financially as an inventor: therefore, the perpetual motion motor is being submitted for copyright. Thank you for your interest in the author's ideas and discoveries. Perhaps one day these ideas and discoveries will serve their useful purposes. No financial progress can be made. The author is sickened by the prospect of not being able to make any financial progress.

Even if the ratio of the sum of the impeller gallery inlet ports areas divided by the sum of the large areas of the four collectors is multiplied to any force and then multiplied to the ratio of the two radii in reciprocal and then one of the two is subtracted from the other while the fulcrum is the axis of rotation of the center gear, still the sum of the moments do not equal zero. Forcing the sum of the moments to equal zero is proving difficult.

Pascal's Laws show that a large area will sweep a short distance a particular volume and the same swept volume will sweep a long distance through a small area. The radii of the two areas in this case are in consideration: the port areas of the meshing gears teeth and the impeller blade vanes. Is the applied force to the impeller blade vanes diminished because the impeller gallery inlet and outlet port areas are smaller than the sum of the meshing gear teeth areas? Where did the applied force go? Was it dissipated somewhere? If so then where is it? If not then the applied force applies at the impeller blade vanes volume, area doesn't matter. Apply the formula for fluid mechanics of inertia, force is applied at the meshing gear teeth to the hydraulic fluid in static elastic equilibrium in zero time present simultaneously with respect to the simultaneous applied force to the impeller blade vanes volume on both radii. Notice how many variables can be changed: 24 can be changed to 16 or 8 or 12, these variables have been mentioned already, and the whole mechanism can be redesigned to accommodate the change in variables. Close the throttle and crank the crank, the perpetual motion motor should start, if it doesn't I don't know why not. Keep trying.

Description of the Super-slipstream Bicycle

Richard Chastain (author) Description Of The Super-Slipstream Bicycle

(Two Wheeled Pedaled Device Powered By A Hydraulic Power Plant)

Starting behind the driver, the first "boxes" are the initial power transmission pumps that pump hydraulic fluid to and from the "box" in the middle in the center on the back in which in the "box" in the middle in the center on the back is an impeller that is to be turned by the hydraulic fluid flow force suction from the first "boxes" behind the driver. The initial power transmission pumps for the super-slipstream bicycle are two quad packs of power transmission units having a gears depth of 13.2 inches.

The paths of least resistance to allow fluid flow to follow around the impeller outer limits (of the second "box") are channels designed in to the impeller on either side of the impeller blade vanes. The force of this hydraulic system is designed on suction and aerodynamics. The hydraulic tubing goes in to and out from the initial pumps (left and right side "boxes" directly behind the driver), on both sides of each pump. The hydraulic fluid lines have bleed valves to purge the air in the lines. A light suction will be applied externally to the suction lines at the bleed valves placed along the suction lines to draw out the air in the suction lines since no reservoir can be placed along the suction lines. The bleed valves will be closed once the air is removed from the suction lines. Hydraulic fluid will be drawn in to the suction lines via the grease seals secondary tubing from the reservoirs through the clearance very slowly at first, a light suction will be applied at the bleed valves to draw out the air in the assembly.

The "cylinders", above and behind the "boxes", have conic sections cut in to them on the inside to funnel the hydraulic fluid in to or out from one line of tubing coming in to the opposite ends of the cylinders from the impeller sides. As the hydraulic tubing reaches the impellers (of which there are two, one is at the rear wheel) the internal diameter of the tubing narrows with respect to the depth of the impeller blade vanes (depth is transverse to the tubing length). This, in effect, creates a velocity change of the hydraulic fluid flow to the impeller blade vanes, which fluid velocity is particular to each impeller. This velocity change is only respective to the upper hydraulic tube of the respective impeller. Two revolutions of the crank with a 24 tooth chainring at the operator develops 16 revolutions at the first impeller and 43 revolutions at the final drive impeller, produced by the hydraulic fluid velocity developed by the first four (quad set of) power transmission units (pumps), then the second set of power transmission pumps drive the rear wheel.

The second pair of power transmission units (left and right rear "boxes") simultaneously deliver hydraulic fluid to and from the second impeller located at the rear wheel. The second pair of power transmission units (secondary power transmission units) have a gears depth of 26 inches.

The [view of the] throttle [is obstructed, it] is [behind a tubing in the view,] next to the rear wheel impeller "box" between the two lines of inflow and outflow tubing. The same principles apply to the hydraulic fluid flow from the second set of power transmission pumps ("boxes") to and from the second impeller only the second impellor is a smaller diameter and the pumps deliver more than 8X the fluid flow volume of the initial pumps with two revolutions of the chainring at the pedals (see formulas) due to the fact that the initial pumps only turn four revolutions per two cranks while the secondary pumps turn 16 revolutions per two cranks of the chainring at the pedals. The result is 43 revolutions per second at the final drive impeller per two crank revolutions of the chainring at the pedals.

A 26 inch diameter wheel turning 43 revolutions per second and a 24 tooth chainring with two revolutions of the chainring in equal time at the crank should require the bicycle to just reach a speed of 200 mph. The wind resistance is estimated to be over 400 pounds, it has not been calculated exactly. Software for Fluid Mechanics Of Inertia is required to calculate the exact air flow force to the vehicle.

Certain measures (tubing wall thickness calculations) have been taken to make sure the tubing does not collapse under the suction load, whether these calculations are correct is as yet untried. As of this moment the only difficulty foreseen is the mangling of the suction tubing under load since the suction tubing will probably all try to become straight as load is applied (this will not happen to the head side of the hydraulic tubing since it suffers no load, only to the suction side). These reservoirs have secondary tubing which feeds hydraulic fluid in to where there is a grease seal as a static seal at each central axis of power transmission unit for each "box" assembly and supplies hydraulic fluid (or anti-freeze and water 50/50 ratio, or sewing machine oil) to prevent the grease seals from sucking air in through the grease seals contact area since there will be a great amount of suction in the gears galleries and impellers housings. The suction force should be minimized because of close tolerances. Hydraulic fluid is supplied to these regions by the reservoirs in both locations which reservoirs' hydraulic fluid is at ambient temperature and pressure and have an opening to prevent draw suction. Once all the air is purged from the assembly the applied force to the hydraulic system should be solid. Bleed valves are placed in strategic locations and a light suction at the bleed valves will disperse trapped air from the main hydraulic tubing. The tubing and power transmission unit casings and plating can be made of polyvinyl-chloride or some other kind of light elastic plastic and their components glued together. Complications of tubing assembly can be remedied with a sleeve and a strategic cut made where the tubing is straight and the sleeve is inserted to cover the cut and the assembly succeeded to by gluing the sleeve to the tubing over the cut. If the entire assembly is made of plastic the tubing can be bonded together with adhesive.

Included in the mechanism are hydraulic brakes with all the required implements. These drawings of hydraulic brakes are not functional as working models in the drawings but are simply representative of hydraulic brakes. Also, there is a twist grip throttle mechanism which throttle is cable and spring operated. The throttle operates by closing off the fluid flow diverted around the final drive impeller so the fluid flow is directed to flow through the impeller gallery: otherwise, with the throttle open the fluid flow is diverted around the final drive impeller and the wheel does not go around. The twist grip mechanism drawing is not functional as a working model in the drawing. These turnkey mechanisms may be purchased separately and are not intended as patentable with the patent application.

The ultimate final drive consists of a connecting chain loop that drives the drive train from the primary impeller input shaft. This assembly has a freewheel at the large chainring which freewheels clockwise (from the left side view) and locks counterclockwise, thus allowing the operator to crank at start-up while the power transmission unit assembly begins functioning. Once the applied apparatuses mathematics take over the ultimate final drive chain loop should apply force to the freewheeling large chainring exceeding the operator's force thereby reducing the operator's applied force to the crank while maintaining and increasing the vehicle's output performance with the throttle while still allowing the operator to crank. This is an unorthodox performance output means but it is mathematically probable and is the basis upon which human powered flight will be developed. This theory is understandable in the mathematics for Fluid Mechanics of Inertia in the power transmission application.

The only three prospects that will prevent this ideal from working up to specifications are suction resistance to fluid flow in the power transmission pump gears, a 90 degree direction change of the fluid flow in the gear pumps, and the fluid flow path in the impeller gallery. The fluid must change direction 90 degrees in the gear pumps twice, although the second time requires no pressure to be applied except to provide its escape. The secondary power transmission must turn 16 revolutions per second at maximum speed, and the primary power transmission turns four revolutions per second at maximum speed. The depth of the secondary power transmission also increases the required fluid flow necessary with respect to the 90 degree change of direction. The primary power transmission is half as deep as the secondary power transmission but there are four times as many of them. Both the primary and secondary power transmissions should apply the same resistance force of suction continuously throughout the instrument's speed range due to the fact that one is twice as deep and turns four times faster and the other is half as deep and turn ¼ as fast but there are four times as many of them.

The fluid flow in the impeller gallery must follow a curved path from the inlet flow area to the draw suction around the impeller. This is constrained by the fluid flow passageways created by the design of channels to either side and above the impeller blade vanes. The remaining clearance of the impeller is limited to only a few thousandths of an inch. The likelihood that the fluid flow will follow the required path around the impellor is probable since it is the path of least resistance.

This instrument is also capable of coasting by which the throttle is released simultaneously (or not) and power is resumed by simultaneously applying cranking and the throttle at speed. The throttle will govern the cadence by regulating the fluid flow volume to pass around the output impeller thus allowing the secondary power transmission to speed up or slow down thereby increasing or decreasing the cadence while maintaining speed.

The operator must consume nutrients in order to apply power.

Names of Human Powered Helicopters:
Twits
Twerps

The Exploits Of Human Powered Flight

The Exploits Of Human Powered Flight

By Richard L. Chastain

The problem with having a gasoline engine, while flying a helicopter is running out of gas and not having a place to land. A human powered helicopter is not like that, the problem with the HPH is you will run out of strength and all you will have to do is set down and wait and your energy will charge back up all by itself. You may also need to eat something but mostly just resting for a little while is enough to catch your breath and get your energy back to further your journey.

The prospects of going after the indigenous psychoactive intoxicant wildlife (i.e. the marijuana plant, the psilocybin mushroom, the peyote cactus (mescaline), the coca tree (cocaine), and whatever else I can't think of) is inevitable. People on human powered helicopters will scour the countryside for these plants and the Earth would be stripped bare of them were there no law enforcement case providing officers with human powered helicopters to prevent such a catastrophe, when actually the law officers will probably use the human powered helicopter to go in search of these indigenous life forms to destroy them themselves. Once the human powered helicopter comes of age mankind will seek out and lay waste to the wilderness of its indigenous psychoactive wildlife. However, this is not a bad thing. Once it is ascertained that the indigenous intoxicant wildlife is in danger of becoming extinct the Government should affect the endangered species act or whatever is necessary to protect the indigenous psycho-active intoxicant wildlife, which is counterintuitive according to the present state of intentions of the present Government. Otherwise they will just wait and let man extinguish the plants from off the face of the Earth. Mankind is not smart enough to think before he acts when he craves these substances, dead or alive, especially in the United States where there is much abundance.

Human powered flight will make it possible to raid farms for food, although it would seem silly to form a raiding party with human powered helicopters since a human powered helicopter will be quite expensive and it doesn't seem likely that anyone who can afford one will be suffering from any kind of starvation prospects and would have no need to form a raiding party: still, in the future raiding parties may occur from natural disaster or some act of God. Farmers will need huge field sized nets to cover their crops so that individuals on human powered helicopters coming to raid their farm cannot swoop down and gather from the farmer's fields. Farmers shall not be allowed to shoot people however, they will only be within their means to call the police. The mesh of the nets will have to be fine enough to keep human predators out of the harvest, cutting through the nets covering a farmer's harvest could probably be a more serious offense, and getting caught one could do time, pay a fine, or both, for cutting a farmer's harvest cover net.

Telephone lines will also be a bother. What will become of the telephone lines and power lines and electrical wires running between the telephone poles in and between cities and towns? It would seem feasible to string a fiber optic line from telephone pole to telephone pole that light up in the evening and stays lit all night so that the locations of the telephone poles and wires are obvious to pilots of human powered helicopters. The event would have to be funded by indiegogo or kickstarter or something like that. The military bases will have to do similarly and string a fiber optic light line around the military base to keep out the non-military public, or think of some other kind of means to deter people from flying in to a military establishment unawares.

The first human powered helicopter will have to obtain permission from Congress to fly over, into, through, upon, and under Federal Land Open To The Public to be able to leave a private restricted area in the first place.

Homes will have their garages built on their roof to accommodate the availability of human powered flight in the future. City skyscraper structures may have landing platforms built outside the offices of individuals for the purpose of convenience of access to one's human powered helicopter and/or office.

Flying this human powered helicopter will be elementary. Considering the modern day traffic or automobiles on interstate highways and during the rush hour, they are all bunched together on a single strip of pavement. Human powered helicopters will free man from the pressures of city life and grounded traffic opening up the sky with all of its innumerable altitudes and directions which can be taken on a whim. Establishing freedom in the sky for mankind in this country will require the Government yielding to the prospect that in time scenic routes will precipitate out and can be adopted as major throughways. For there to be determination to control the trajectories of our flights by imposing on man to dress and cover in flight, or to fly a designated flight path laid down before a scenic route has precipitated out, or to cause single file flying, bottlenecking et cetera, will need to be outlawed or prohibited in some way thereby eliminating traffic jams and a major cause of accidents in the sky. The deregulation of human powered flight should be a Right (as is prohibition) established by Government to free the people from catastrophe and so that our wings can never be taken away from us by any justification of justice. Owning a human powered helicopter should be a Right under the Fourth Amendment, as a keeping and bearing of arms and a rule needs to be applied that prevents the police from the confiscation of a human powered helicopter as a weapon is confiscated during an arrest.

As the nobleman goes along on his way, so the nobleman shall be free on his journey according to the Scriptures of the Holy Bible. Traveling in to wildernesses on a human powered helicopter should not be an excuse for law enforcement to pursue after the individual with the intent of apprehending them on any grounds without any probable cause, preponderance of evidence, burden of proof, or eye witness of a crime having been committed by the operator of the human powered helicopter only by the suspicion that an operator of a human powered aircraft is a drug addict. Appearing to not be doing anything does not justify apprehending someone in a National Park and imposing on the person to have to serve time in a mental hospital because he looked like he wasn't doing anything, in the National Park, should be investigated and set right. Since when does appearing to be doing nothing justify imprisonment in a mental hospital? Occupying one's free time in a National Park by one's self is not a crime, nor is it blatantly insane, or any intent to commit a crime and does not justify so much as even an approach by a law enforcement officer or Park Ranger. Flying to the mesa of a chimney in a human powered helicopter out West would deter any such approach, and sit there and appear to be doing nothing for the rest of the afternoon, you probably won't get approached by a cop on the ground up there and being taken in to custody and carried off from the top of a mesa tower, to serve time in a mental hospital, may prove difficult.

The objective by Government securing the sky as a National Preserve renders the requirement of a radio transceiver with multiple transmitter power settings which can be used to communicate short distances as well as long distances by the flip of a switch without shouting across the sky to another operator or shouting at all, because it is silent up in the sky except for the wind blowing and rain and thunder and the only shouting to date is the birds shouting to one another, birds are always shouting anyway, and the preservation of the silence should be kept secure, the sky should be kept as wild for future generations as it is now. Painting the clouds with food coloring should also be restricted to special occasions and the clean water act should take a precedent in case of overenthusiastic cloud painters making the water taste bad eventually, and change color from clear to a mucky brown from the mixture of all the different colors of food coloring.

It will be impossible for the operator of one human powered helicopter to reach out and touch the operator of another human powered helicopter, as you can determine by the allowances of the blades locations during flight when approaching another human powered helicopter.

Still, it is not a wise decision to sow seeds while aloft in a human powered helicopter unless farming on one's own land, with the exception of sowing seed balls. Leave sowing wild seeds to the birds. Birds love to eat seeds and expel seeds, and there would be plenty of seeds to be sown if mankind had not got it in his head to lay waste to the indigenous foliage life forms over all the Earth in the meantime, on his human powered helicopter. Sit there on your sack of seeds and just smile and wave. Then there is the issue of where to land: in National Parks manmade "trees" can be built in remote places that will support human powered helicopters as landing pads and instead of landing on the earth and causing any erosion they can land on the "trees" provided for them: otherwise, it probably would be best to land on rock or concrete to prevent soil erosion in the wilderness, despite the fact that the human powered helicopters will probably only be able to land on flat surfaces including the water if so equipped. These "trees" will be made of materials and construction designs to support human powered helicopters like landing on fixed and rigid leaves. With pontoons you will be able to go fishing, and fly home to your lovely family and fix what you caught to eat.

Getting chased by the cops will change too. From once long ago there was running and trying to get away being pinned to the Earth by gravity. Now there will truly be flight, and fight or flight will change to fight and flight, and eluding pursuit will be something to deal with. Dog fights may erupt on rare occasion between the assailants and the police.

The Holy Bible makes a reference to a holy multitude coming so dense that the light of the sun was darkened and the sound of their wings was as the voice of God. Should it be possible for everyone in this country to own a human powered helicopter then a mighty fighting force will we be in the event of belligerence, should the military need assistance. As well, human powered flight will make mankind capable of migration in the event of a solar system-wide cataclysm.

There should also be outlawed shooting wildlife from a human powered helicopter, including coyotes. The author does not think anyone should shoot any animal from an aircraft, how unfair is that. The author believes hunting from an aircraft is already outlawed in the United States but I don't know for sure. The author has seen a video of shooting coyotes from ultra-lights or something like that years ago on TV. The fact is the author cannot even fly the human powered helicopter that he would build even three inches off the ground without breaking the law these days. His human powered helicopter design would be an unregistered and uncertified aircraft in the United States' air space once it was built and would be subject to confiscation and he would get six months in jail if he allowed myself to be caught flying his human powered helicopter in the United States off of private property. Which is where deregulation of human powered flight is adamant, and human powered helicopters should be seen and privileged as an environmentally non-invasive contrivance of aircraft capable of sustained flight, and not requiring a license to operate. In which case you will probably need a good life insurance policy because these are the beginnings or the 21st Century, not the 22nd Century, and human powered helicopters are just about to be born in to this world and one will not be the most technologically advanced new thing since the model-T 100 years ago was not a speedy Indy race car of the 21st Century, so give leeway to the effect that 100 years have to pass before there will be the most technologically advanced human powered helicopter the world has ever seen.

Up the chimney will be a game: fly to a high altitude and cease cranking, let the human powered helicopter plummet to the Earth, and with precise enough timing engage cranking just in time to slow to a stop and touch the ground with the landing gear. Children will get good at this game. It is like driving the tractor when cutting the grass, practice, practice, practice.

In the most probability the favorite altitude will probably be to fly just above the tree-tops. This will facilitate fewer injuries from falls should a human powered helicopter fail to perform for some reason. A fall in to the trees may cause less serious injuries, and falls from low altitude as well will cause fewer serious injuries.

Human powered helicopters may not be user serviceable except for greasing bearings, adding hydraulic fluid, making minor changes to the operator's compartment or changing a chain, and some more serious replacements of some parts. Most of the human powered helicopter is not user serviceable: however, the human powered helicopter is designed to be maintenance free and should give more than a lifetime of good service, with proper care.

Human powered helicopters also present the formidable capability of spying on groups or individuals. Going in seeking of "where it's at" is also a formidable prospect for the use of a human powered helicopter. Catching a boyfriend or girlfriend cheating will be a capability of human powered helicopters. The flocking together of individuals will occur. It will at once be possible to see soul mates from a distance in the air, and flocks of people may form in time. This will also facilitate that these individuals will accumulate in neighborhoods as companions, or extended "families". Extended families, akin to flocks of birds, would more probably be the wiser organization of flocks of people rather than organizations of large flocks of "friends". Flocks of "friends" would be more likely to be disorganized and leave a trail of infractions in their wake.

Once the neighborhood has been established, then it will be possible to design cities and towns or communities around the human powered helicopter, with spokes of taxiways radiating out from a central runway. Houses, homes, cities etc. may be built on these taxiways: however, if HTOL human powered aircraft are not employed then with landing platforms on the roofs of houses there may be no need for taxiways or a runway for human powered helicopters. It may even be possible to have a job in another city should the human powered flight prospect prove fast enough to get to work with if one lives in another town from their workplace. However, in the meantime parking will be at a premium for human powered helicopters in parking lots.

Celebrities will have to carry umbrellas if they don't want to be recognized from the air. Nudist colonies will have to cover up. One day there will be a law that you cannot fly a human powered helicopter unless one flies wearing no clothes, but then this may be in extreme circumstances of the necessity to have to re-populate the planet should the human population drop to, say, twelve individuals or some critical limit where people are dying faster than they are being born and the population will become extinct in a matter of, say, half a generation. No one can tell the true circumstances of the Revelation To John unless they are Saved by The Grace Of God and can truly interpret the scriptures with reality. Who knows? Maybe there will be a time in the future when the population of the Earth is in danger of becoming extinct in a short time. Until then, this wearing-no-clothes business is just foolishness for me. Perhaps it is true, but how can I tell the future?

Nevertheless, putting restrictions on human powered flight, certainly for human powered helicopters, will only embitter the population, like trying to make them stop drinking coffee, with some sort of jurisdictional restriction.

The human powered helicopters should come equipped with a GPS, an altimeter, a turn and bank indicator, and a compass, and a two-way radio all incorporated in to a 12 volt DC electrical system with an alternator for power and all the necessary implements of 12 volt DC electrical systems with all loads running in parallel. This should be sufficient equipment to operate a human powered helicopter in Federal airspace open to the public. Once again these human powered helicopters are designed to be pilot-license-free and deregulated so mankind can enjoy the freedom of the skies without the burden of regulations and be free as birds to fly in the common air, unhindered by any persistent influence of justice during their traverse.

Speaking of which, there is the invasion of privacy issue, should one appear in the sky where another is in anticipation of their privacy being kept confident: therefore, privacy (in Kentucky) is lawfully maintained at an arm's length-to be to the tips of one's fingers as far as one can reach, and should there be any dispute as to whether one is "invading one's privacy" by appearing somewhere over the horizon, it shall be noted that in Kentucky privacy is limited to an arm's length and should any dispute arise from a "spy" anticipating the capturing of views, photographs, or whatever kind of media a record can be made on, of some individual, group et cetera, so as to ascertain facts-on-the-ground that the limits of privacy do not extend "as far as one can see" but to the tips of ones fingers as far as one can reach. Therefore privacy does not extend in to space as far as one can see, nor does ones property line extend vertically to the outermost reaches, but it is only an arm's length to the tips of one's fingers as far as one can reach. One should then be able to come in to view, approach, land and walk up to the suspected subject, or suspect, and stay just outside the full reach of their fingertips and NOT BE INVADING THEIR PRIVACY and take pictures, make recordings, or just look at them, just to get the wisdom and understanding that one should choose otherwise another mate, gather information or whatever is required. As long as one does not come inside of the arm's length extenuation then privacy is not invaded. This arm's length law needs to be established throughout this country in the event of human powered helicopters and the capability of being able to ascertain wisdom and understanding in the light of one's relationship with another. Breaking off a relationship on the pretense that you succeeded at finding your mate cheating, because you have a human powered helicopter and it gave you the capability to spy on them, is every reason to establish an arm's length privacy law nationwide and to outlaw privacy being "as far as one can see" (in any direction), because the mate will use the defense "as far as I can tell, or as far as I can see, or as far as I can understand" or such like, to establish their privacy in the courts, and the judge will need to establish that the "as far as..."rule, anything when it comes to privacy is that the realm is limited to within the tips of one's fingers as far as one can reach: as well, preferential treatment (or mistreatment in the case of a male defense witness) for the female prosecution witness by judges, and the defense council, in courts needs to be outlawed with fines or imprisonment or both and should be investigated.

Vulcanization of property to human powered helicopters should be limited to any three points on the property, whether it be a building, growing structures, vehicles, fences, persons et cetera, and the lines that join the three points creating an assembly of areas covering the entire property inside the property line. And the consolidation of properties between neighbors to increase the size of the vulcanized perimeter should be restricted.

Considering the prospects of one stealing a human powered helicopter: once before from the previous design it was not possible because the device was constructed to tailored tolerances for each individual, but now the aircraft is sustainable in flight by articulation of the blade vanes and some variation in operators weights can be accommodated for. So it may be possible for someone to manage to perpetrate theft of a human powered helicopter and make off with one if their weight is not too significantly different from the owner's. The human powered helicopter will need some kind of lock.

This human powered helicopter would not be too purposeful for use in the military although it may be useful for moving troops in large numbers over great distances in a shorter period of time than presently accommodated for, it is vulnerable to attack since the powerplants are made of Styrofoam and epoxy-carbon fiber. The operator is exposed to raw gunfire and has virtually no protection besides body armor. This aircraft may prosper for military means some distance behind the front.

There is also the prospect from the Holy Bible that the high places shall be made low and the valleys shall be raised up even with the plains. This may be a reference to human powered flight.

As for farming, the human powered helicopter may be used to lift somewhat heavy objects and carry them distance, or do some heavy lifting of sorts. Also, as with minor heavy lifting the military or Department of Transportation may be able to use the human powered helicopter to go on search and rescue operations. The human powered helicopter may also be used as a stealth aircraft because it may be very quiet. The Corpman may have use of the human powered helicopter in combat, with the help of a GPS and the coordinates of injured soldiers.

Going camping will now be a strategy curse for the camper-outer. The police will be like Were-wolves on human powered helicopter campers because of their likelihood of carrying illegal drugs despite the fantastic corporate reality that human powered helicopters will cost probably $280,000 for a new one easily and that casual consumers of controlled substances won't be able to go human powered helicopter camping because they won't own one. But who will go human powered helicopter camping anyway? Human powered helicopters are designed with carrying capacity now and practical camping is within the means of human powered flight. So beware of the police cracking down on you for using drugs if you have a human powered helicopter. Be aware thou owner of thine human powered helicopter, thou art a drug addict, and the police are going to make you know this, so be prepared to be searched and taken into custody if you are brought down by a police officer, and to spend the night in jail and to have your human powered helicopter impounded or confiscated. Therefore, the Rebellion: The Rights of the People to Possess, Own, and Operate human powered helicopters in Federal airspace and between the ground and their surface ceiling without fear or interference from law enforcement in that an individual shall be free on their journey (this implicitly includes California, without exception and without exclusionary rules applying, California is to be disciplined in its mistreatment of interstate traffic and corrected). Also, The Smokey Mountains are to be managed by Federal Authorities in the careless mistreatment of interstate traffic and corrected where human powered flight is prejudged as a means of trafficking in controlled substances or simply as a means of transportation for a drug addict when an individual on a human powered helicopter is seeking to engage in wilderness adventure.

Any location of public interest, mainly national parks and forests, recreation areas et cetera, where individuals are found to be being mistreated having been perpetrated for offenders for being occupants of human powered helicopters will be investigated and the suspected department of law enforcement subject to disciplinary action and corrected if there were no charges for drug offenses and all the subjected offenses were a result of the officer(s) having apprehended the subject(s) and all the offenses were ex post facto, e.g. concealed deadly weapon, public intoxication, assaulting a police officer, resisting arrest, none of which are drug related, and all are after the fact, since the police office arriving on the scene is more than likely after an individual on a human powered helicopter for drugs than for a pocket knife, or for having had one beer, or for talking with their hands, or for struggling because the officer just broke their arm having had them in a hammer lock. Nevertheless, the fact remains that the motive for the mistreatment has to have been drugs related, and finding out whether or not there are any drugs during the scene during the arrest is paramount in the discovery of the motive during the investigation. No police officer shall be left unturned.

If you leave the human powered helicopter alone long enough spiders will make their webs in it, birds will make their nests in it, the paint will flake off, it'll rust, it'll rot, it will leak, it'll get wet in the rain, snow will collect on it and ice will hang down off of it. It'll collect dust. The battery will go dead, branches will fall on it. Mud will get splashed on it. Parts of it will fall off or get stolen. It will sink in to the earth and begin to lean over. Eventually it will fall over and no one will pick it up neither can it pick itself up, and after a long enough time it will turn to dust. It is not a God. It cannot see or hear or speak or work miracles. It attracts lightning it does not send it forth. It can be broken into pieces and smashed to smithereens. If you ask it a question it does not respond because it cannot answer. It is lifeless and does not even know what it is. It does not know anything. It is a mechanism, a machine, which man operates with his mind and body. It itself can do nothing. If something breaks on it while you are in flight you could die, it cannot save you it has no powers of self-preservation. Do not worship it. It is made by craftsmanship as are all machines. Human powered helicopters are not Gods, or God, or any God. They are machines and are to be cared for as machines and not neglected lest they fall in to disrepair for they age as mankind ages because of the imperfection of sin and their days are numbered. This human powered helicopter is however a manifestation of The Will of the Christian God (I am), Jesus the Son of God, and The Holy Spirit having used a mere human to work as His conduit for doing His Will. The Human Powered Helicopter however is not yet manifested in physical form, only in design concept reflecting that mankind is not ready for human powered flight at this time, but soon. The fact that the author is financially unable to accomplish the construction of a human powered helicopter leads him to believe that he has much research to perform still and many months of study as well because there are still lots of loose ends in the development of a practical human powered helicopter.

Design Parameters

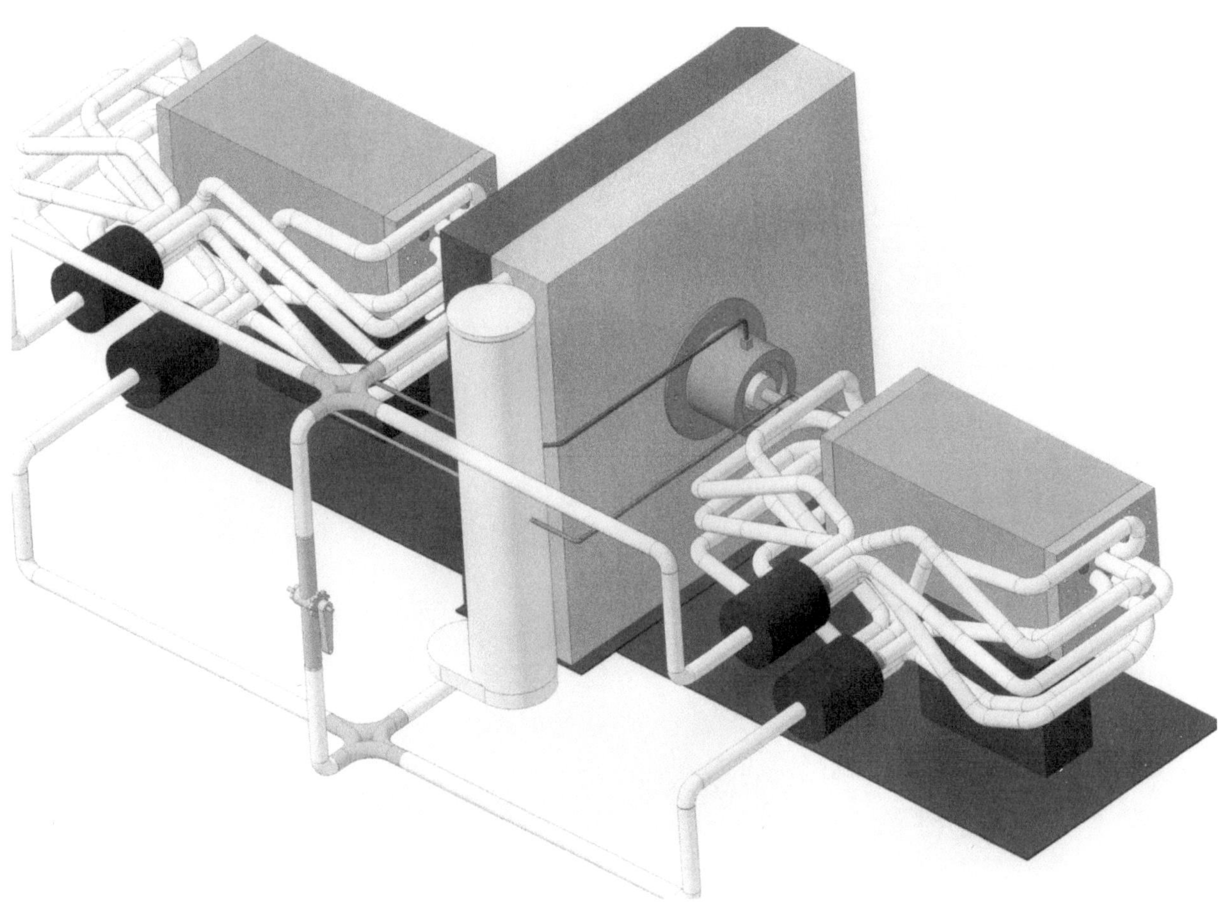

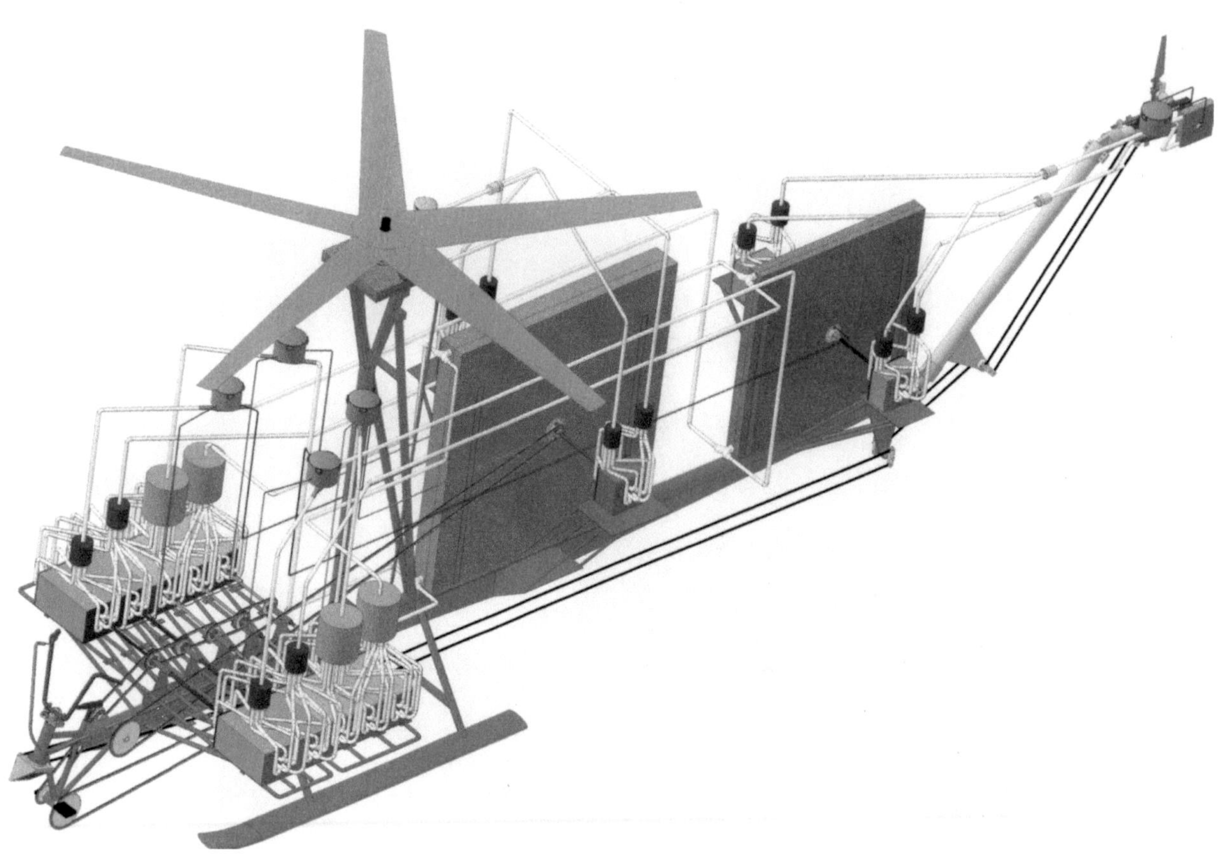

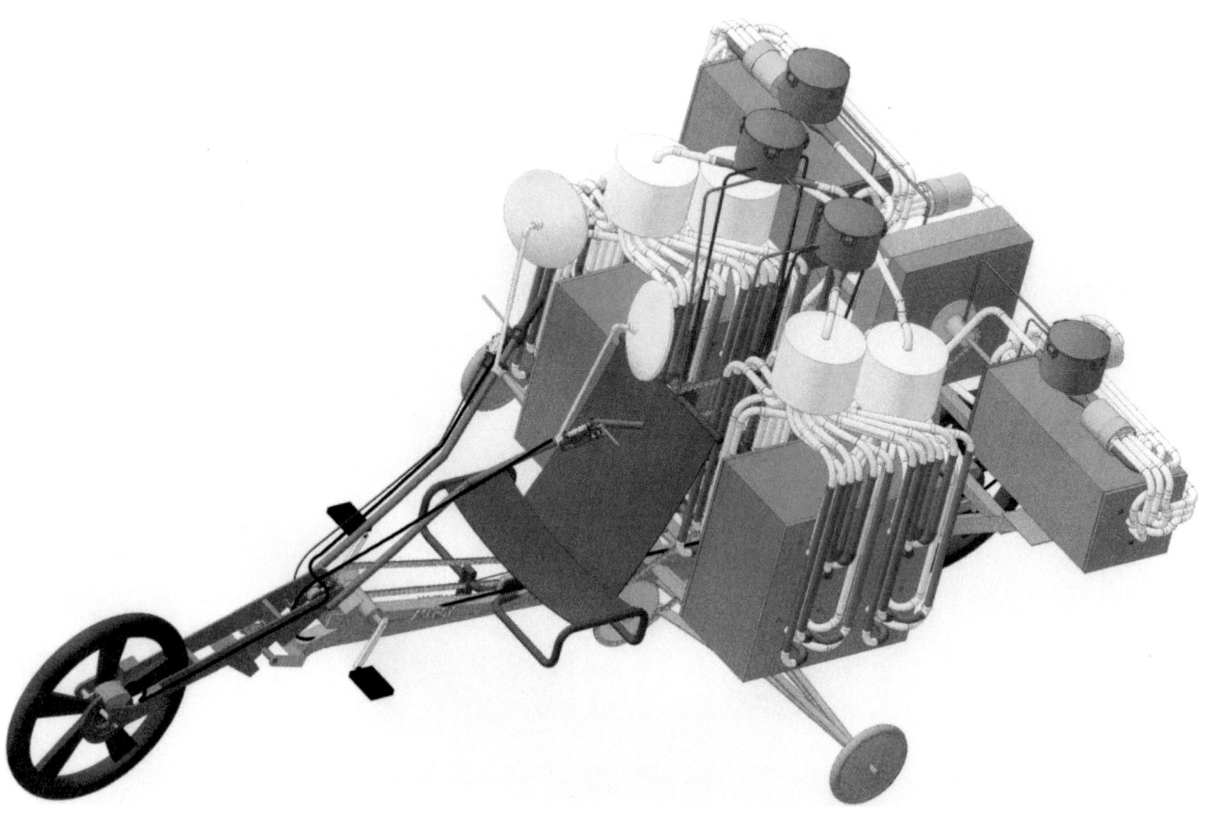

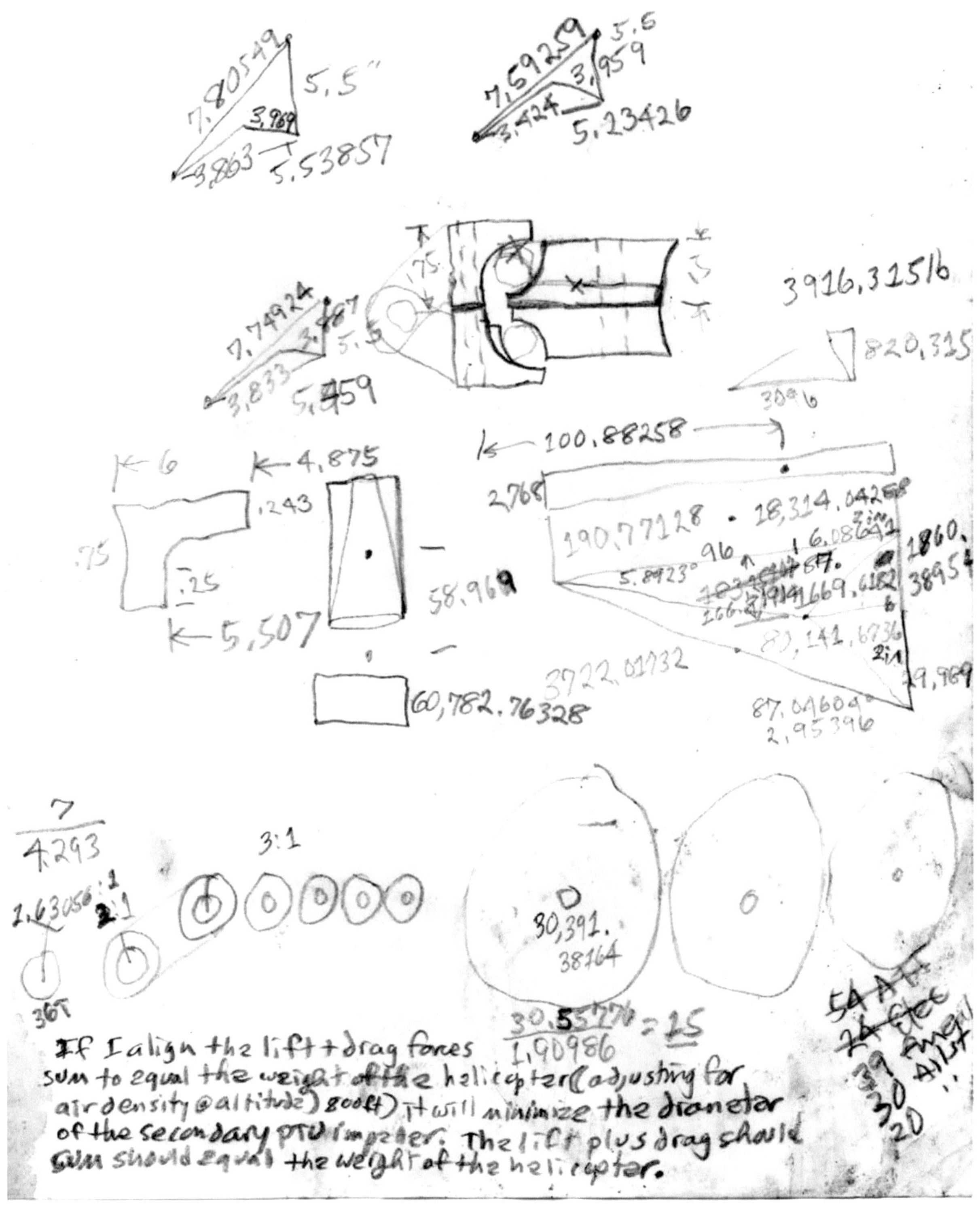
If I align the lift + drag forces
sum to equal the weight of the helicopter (adjusting for
air density @ altitude 800ft) it will minimize the diameter
of the secondary PTU impeler. The lift plus drag should
sum should equal the weight of the helicopter.

2048
14.84°
20°
524.537 (1)
B
1979.688
20°
(3.77416)
A
= 2504.225

Make "2048 lbs" equal to the weight of the helicopter at 84 cadence.
The hypotenuse
I think this is a hover evaluation

A+B = 2048

2504.225 lbs of applied force to the lift impeller will hover 2048 lbs of helicopter at the respective air density and lift blade rpm %. So the helicopter need not weigh 2048 lbs; however, the 84 cadence is still maintained.

$\frac{2504.225}{2048} = 1.22277$, $\frac{3.77416}{1.22277} = 3.08658$, 1979.688 ÷ 3.08658 = error.

3.77416 × 1.22277 = 4.61493, 1979.688 ÷ 4.61493 = 428.97469 lbs

error
$\sqrt{(2048^2 - 428.97469^2)}$

2002.56953 + 428.97469 = 2431.54422 error

2048 − 428.97469 = 1,619.02531 lbs

hypotenuse = 1674.89171 lbs of aircraft.

A+B = 1,619.02531 + 428.97469 = 2048 lbs.

$\tan^{-1}\left(\frac{428.97469}{1619.02531}\right) = 14.84°$

There may be just a little range on a graph where the weight of the helicopter at the lift blade vanes output performance resultant vector and the actual weight of the helicopter coincide.

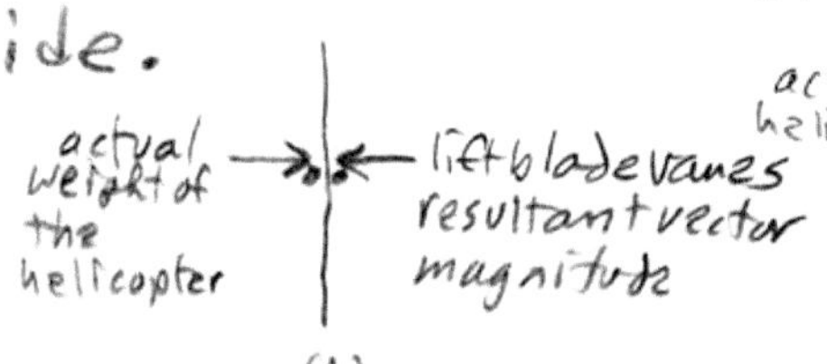

actual weight of the helicopter → ← lift blade vanes resultant vector magnitude

(A)

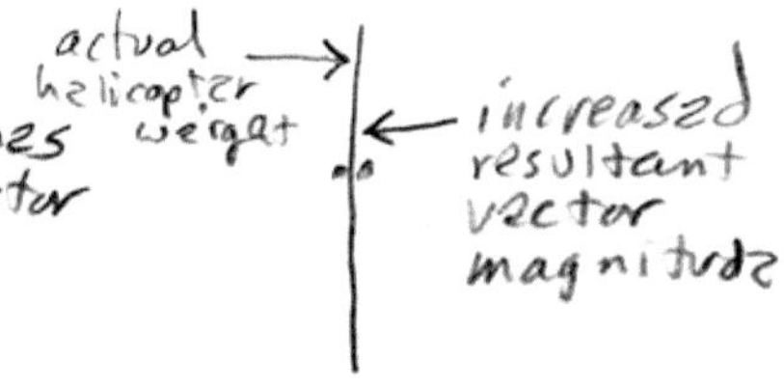

actual helicopter weight → ← increased resultant vector magnitude

(B)

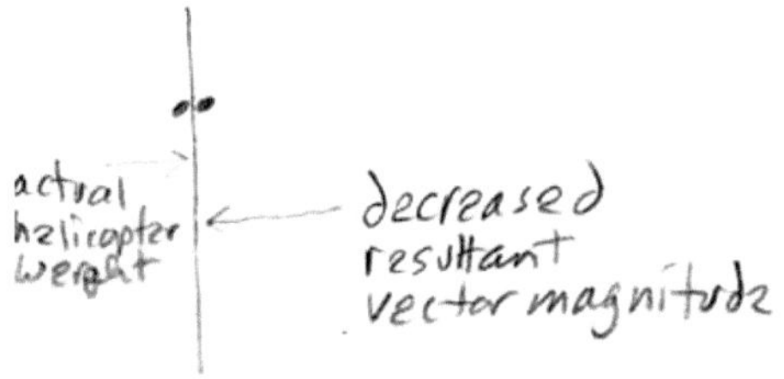

actual helicopter weight ← decreased resultant vector magnitude

(C)

(A) is optimum minimum.

(B) is increasing the lift blade performance but the weight of the helicopter goes up too much.

(C) past point of optimum minimum reducing the weight of the helicopter further bring down the optimum minimum and loss of power is the result. Trimming weight now sacrifices structural rigidity; otherwise, weight must be taken from the power transmission assembly.

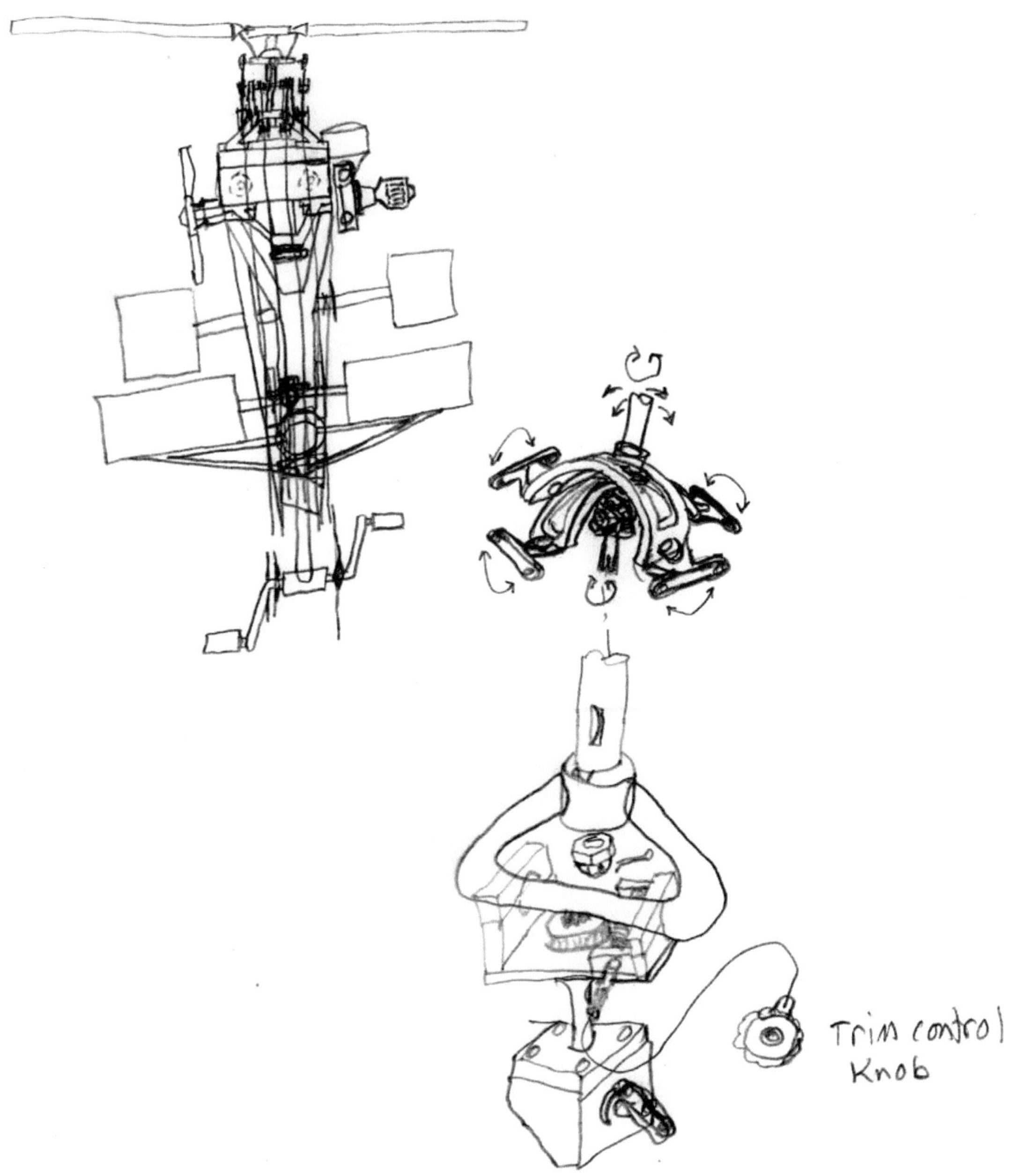
Trim control
Knob

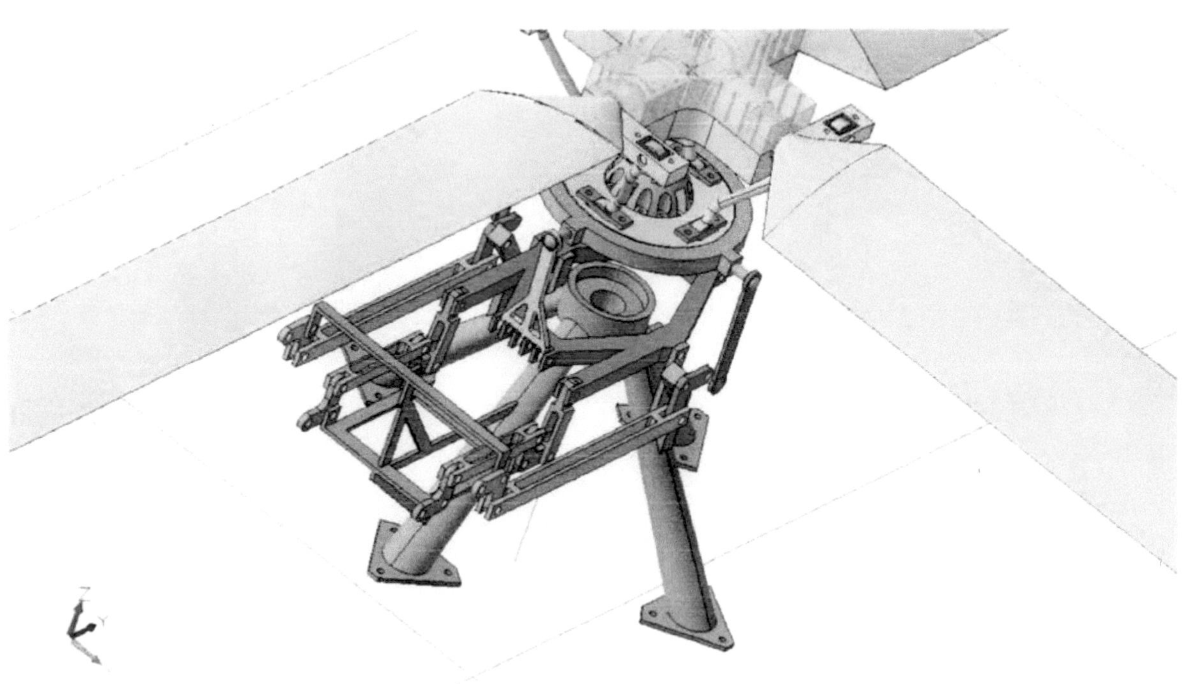

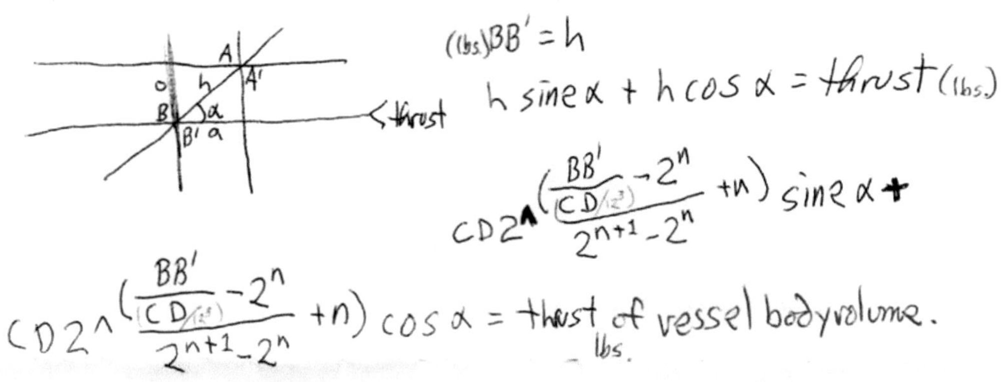

$(lbs)BB' = h$

$h \sin e\,\alpha + h \cos \alpha = thrust\,(lbs.)$

$$CD2^{\wedge}\left(\frac{\frac{BB'}{CD/(12^3)} - 2^n}{2^{n+1} - 2^n} + n\right) \sin e\,\alpha +$$

$$CD2^{\wedge}\left(\frac{\frac{BB'}{CD/(12^3)} - 2^n}{2^{n+1} - 2^n} + n\right) \cos \alpha = \text{thrust}_{lbs.} \text{ of vessel bodyvolume.}$$

The necessity to have to calculate all the vectors for all the subsegments is negligible; all you will have to evaluate is the final result from the geometry of the entire vessel body BB' to AA' with respect to the angle α resulting from the equal volumes opposite perpendicular at AA' and BB' (Solid body geometry).

Multiply by thrust $\left(\frac{\frac{\text{Vessel body volume}}{\text{blade vanes volume}} - 2^n}{2^{n+1} - 2^n} + n\right)$ ~~× blade vane volume~~

$$C\cdot bvv\,2\left(\frac{\frac{\text{thrust lbs}}{(C\cdot \text{blade vanes volume}/12^3)} - 2^n}{2^{n+1} - 2^n} + n\right) \cos\beta + C\cdot bvv\,2\left(\frac{\frac{\text{thrust lbs}}{(C\cdot bvv/12^3)} - 2^n}{2^{n+1} - 2^n} + n\right) \sin e\,\beta$$

= applied engine performance force at blade vane geometric center.

Propeller blade vanes volume

C = air density, bvv in cubic inches
BB' is in pounds.
Thrust is in pounds.

$$C\cdot bvv\,2\left(\frac{\frac{BB'}{(C\cdot bvv/12^3)} - 2^n}{2^{n+1} - 2^n} + n\right) \cos\alpha \cos\beta + C\cdot bvv\,2\left(\frac{\frac{BB'}{(C\cdot bvv/12^3)} - 2^n}{2^{n+1} - 2^n} + n\right) \sin e\,\alpha\, \sin e\,\beta = \text{thrust}$$

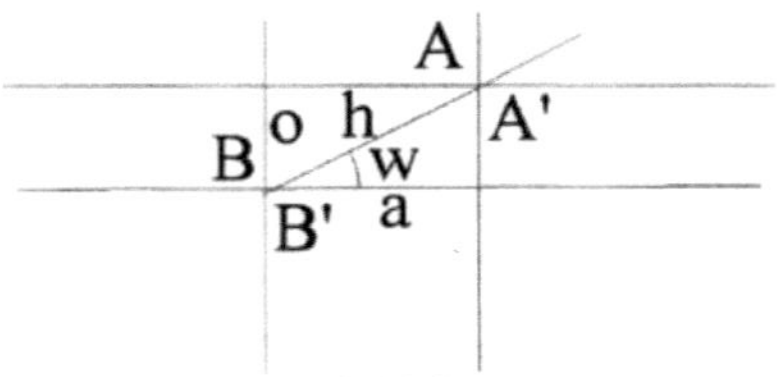

A
B o h w A'
B' a

(lbs.) BB' = h

h sine w + h cos w = thrust (lbs.)

$$CD2\wedge\left(\frac{\frac{BB'}{(CD/12^3)} - 2^n}{2^{n+1} - 2^n} + n\right) \text{ sine w } +$$

$$CD2\wedge\left(\frac{\frac{BB'}{(CD/12^3)} - 2^n}{2^{n+1} - 2^n} + n\right) \text{ cos w} = \text{thrust (lbs.) of vessel body volume.}$$

The necessity to have to calculate all the vectors for all the subsegments is negligible; all you will have to evaluate is the final result from the geometry of the entire vessel body BB' to AA' with respect to the angle w resulting from the equal volumes opposite perpendicular at AA' and BB' (Solid body geometry).

$$Cxbvv\ 2\wedge\left(\frac{\frac{BB'}{(C \text{ x } bvv / 12^3)} - 2^n}{2^{n+1} - 2^n} + n\right) \text{ cos w cos v } +$$

$$\text{Cxbvv } 2\hat{}\ \frac{\left(\dfrac{BB'}{(C \times bvv / 12^3)} - 2^n\right.}{2^{n+1} - 2^n} \left. + n\ \right) \text{ sine w sine v} = \text{thrust}$$

bvv = blade vanes volume of turbine blade or propeller in cubic inches
C = air density
BB' is in pounds
Thrust is in pounds

Sine v is the angle of association made with the first thrust of vessel body volume made with angle w applied to the blade vanes hypotenuse and its adjacent side. The original thrust of vessel body volume applies to the hypotenuse of the blade vanes applied load of what would be WR (wind resistance) bearing on the aircraft at h sine w + h cos w = thrust (lbs.).

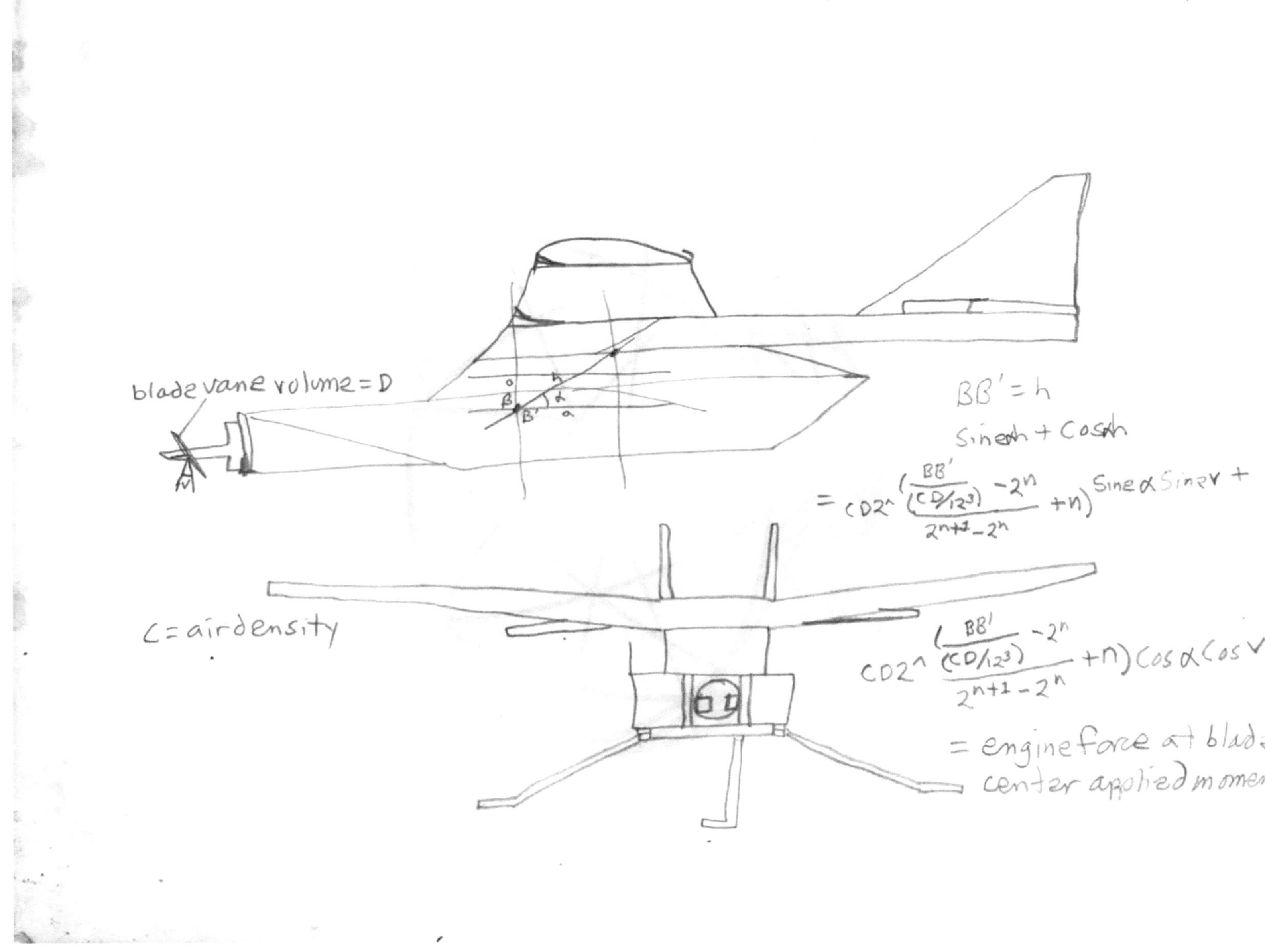
blade vane volume = D
BB' = h
Sine h + Cos h
= CD2^ ((BB'/(CD/12^3)) - 2^n)/(2^n+1 - 2^n) + n) Sine α Sine V +
C = air density
CD2^ ((BB'/(CD/12^3)) - 2^n)/(2^n+1 - 2^n) + n) Cos α Cos V
= engine force at blade center applied moment.

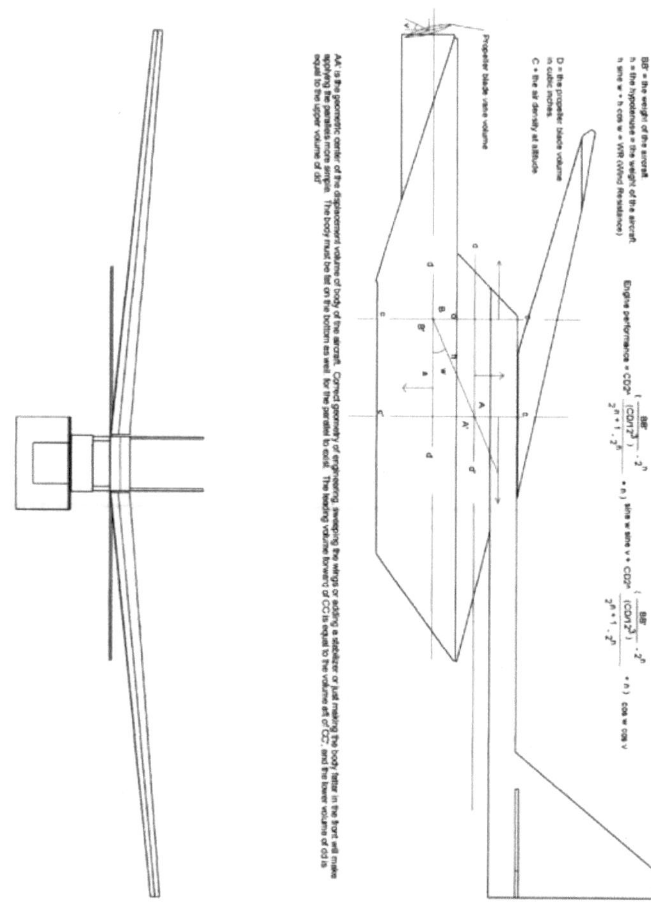
BB' = the weight of the aircraft.
h = the hypotenuse = the weight of the aircraft.
h sine w + h cos w = WR (Wind Resistance)
Engine performance = CD2^((BB' / (CD/12^3)) · 2^n / (2^(n + 1) · 2^n) + n) sine w sine v + CD2^((BB' / (CD/12^3)) · 2^n / (2^(n + 1) · 2^n) + n) cos w cos v
D = the propeller blade volume
in cubic inches.
C = the air density at altitude
Propeller blade vane volume
A
A'
B
B'
w
h
c
c'
d
d'
AA' is the geometric center of the displacement volume of body of the aircraft. Correct geometry of engineering, sweeping the wings or adding a stabilizer or just making the body fatter in the front will make applying the parallels more simple. The body must be flat on the bottom as well, for the parallel to exist. The leading volume forward of CC is equal to the volume aft of CC', and the lower volume of dd is equal to the upper volume of dd'

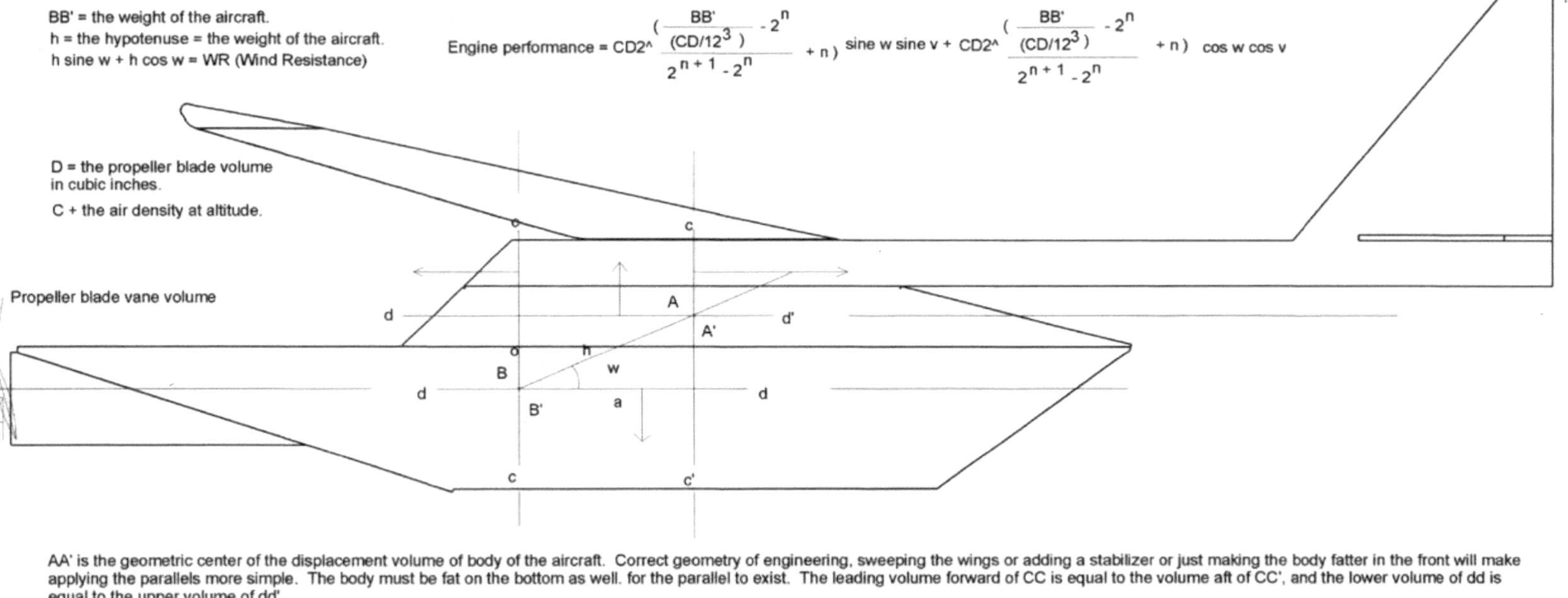

AA' is the geometric center of the displacement volume of body of the aircraft. Correct geometry of engineering, sweeping the wings or adding a stabilizer or just making the body fatter in the front will make applying the parallels more simple. The body must be fat on the bottom as well. for the parallel to exist. The leading volume forward of CC is equal to the volume aft of CC', and the lower volume of dd is equal to the upper volume of dd'.

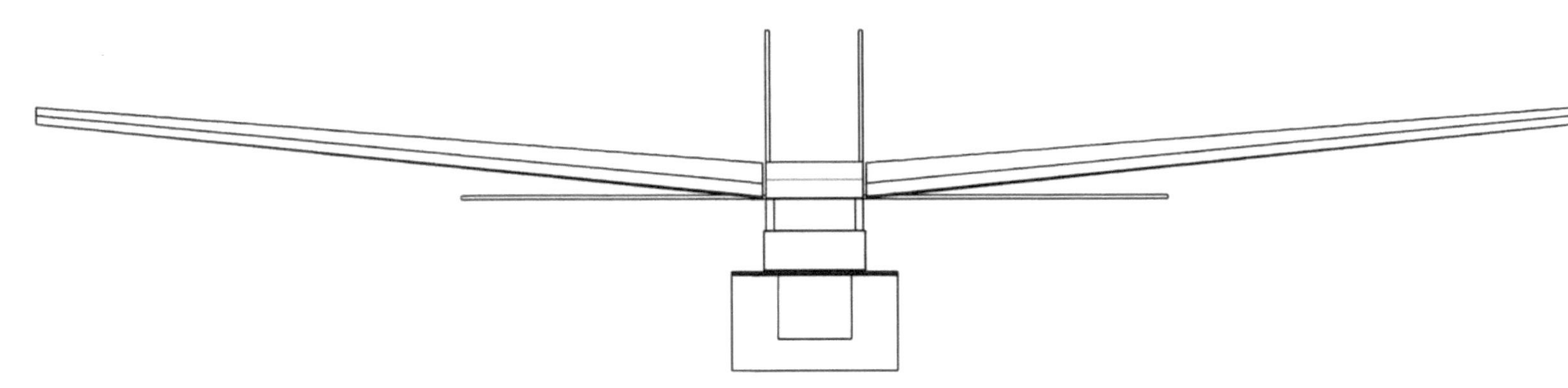

Then there's the impact reaction force(s) reacting on both BB' and AA' at velocity in equal and opposite reaction to impact 100% elastic when applied to F, and $h \sin\alpha + h \cos\alpha$ = Thrust = F. Knowing the velocity where BB' = 1 impact reaction force, as well where AA' = 1 impact reaction force has not yet been identified, and tables of common items are not yet made.

$-FC = -h\cos\alpha e + h\sin\alpha f$

$F\ell = \frac{+h\cos\alpha e - h\sin\alpha f}{C}$

e = 1
d = 9
c = 3
f = 6
g =
i =

$\Sigma m = 0$

h sin α + h cos α

h sin α

$+h\sin\alpha g - h\cos\alpha i + BB'd = \emptyset$ solve for

$\Sigma m = \emptyset$ @ F

$+h\cos\alpha e - h\sin\alpha f - FC = \emptyset$

$BB'\sin\alpha f + BB'\cos\alpha e - FC$

$BB(-\sin\alpha f + \cos\alpha e) - Fc$

$-\sin\alpha f + \cos\alpha e = \frac{Fc}{BB}$

Booboo-error over 500 billion

$-F_i + BB'f = \emptyset$ @ AA'

WR

Solve for F.

$\frac{BB'f}{i} = F$

In this case F = 3, $\Sigma m = \emptyset$ @ AA'.

Ground-hugging

$\frac{(h\sin\alpha + h\cos\alpha)c}{d} = BB'$

$F = h\sin\alpha + h\cos\alpha < \frac{Fc}{d} = BB'$

$BBd = -h\sin\alpha g + h\cos\alpha i$

$BBd + h\sin\alpha g = h\cos\alpha i$

$BB' = \frac{h\cos\alpha i - h\sin\alpha g}{d}$

Workingclassfoodies.com

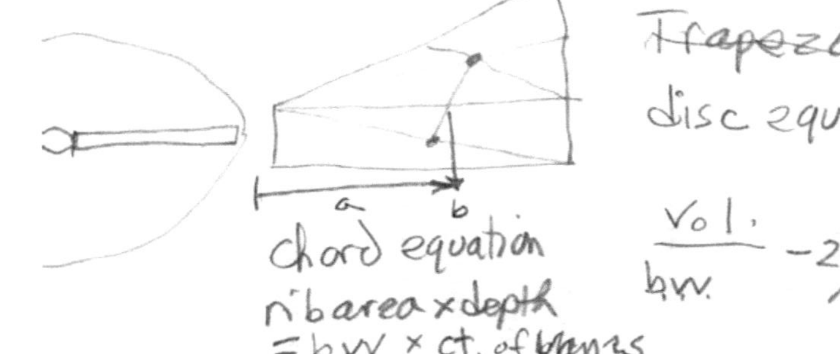

~~Trapezoid~~

disc equation $\pi r_1^2 h_1 - \pi r_2^2 h_1 = \text{Vol.}$

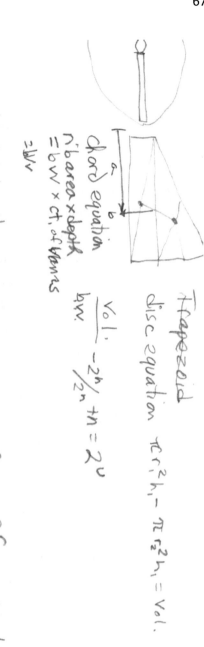

$$\frac{\text{Vol.}}{\text{bvv}} - 2^n/2^n + n = 2^v$$

$$\frac{\text{weight of helicopter}}{2\left(.07651\ \frac{\text{lb/cf}}{12^3} \cdot \text{bvv}\right)} - 2^n/2^n + n = 2^q,\ 2^q/2^v = 2^r,\ r = \text{revolutions}$$

weight of h

weight of helicopter x cos w + weight of helicopter x sine w = applied force at b

The radius of the lift blade impeller is $\frac{a}{8}$.

applied force at b x 8 = applied force at lift impellor blade vanes.

$$\left[\frac{a}{8} = 1\right] \quad \frac{\left(\frac{a}{8}\right) \times (\text{applied force at } b \times 8)}{\left(\text{distance to tail rotor axis} \big/ \left[\frac{a}{8}\right]\right)} = \text{applied tail rotor force.}$$

Calculations for the tail rotor blade are the same as for the lift blade. The human powered helicopter requires two sets of power transmissions, one to drive the lift blade and one to drive the tail rotor blade.

B+ = Bicycle cornering G = gravity

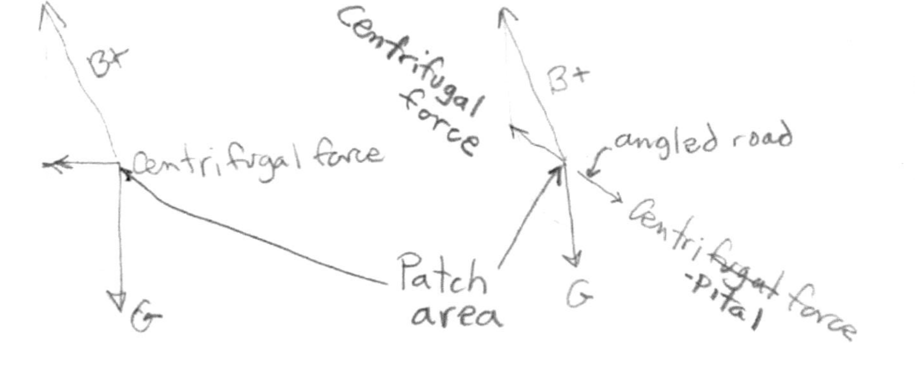

as long as the vectors remain static the vectors directions are against the applied forces. Centrifugal force is in compression so its vectors face one another.

Vehicular force
5x
157.1882
7.81779
inclined road
.57358 x
centrifugal force
4.16085
.57358
.81915
145°
55
α
G
1x 35°
2.18823
7.81779°
4.22001 x
centriptal force
Patch area

~~Vehicular force · cos α +~~ Calculations show the centrifugal force of a 35° incline will exact .57358 G's at 5 G's applied force against the pavement with 1G load.

x = G's

~~Light up the darkness~~ ~~Hanover Saints~~

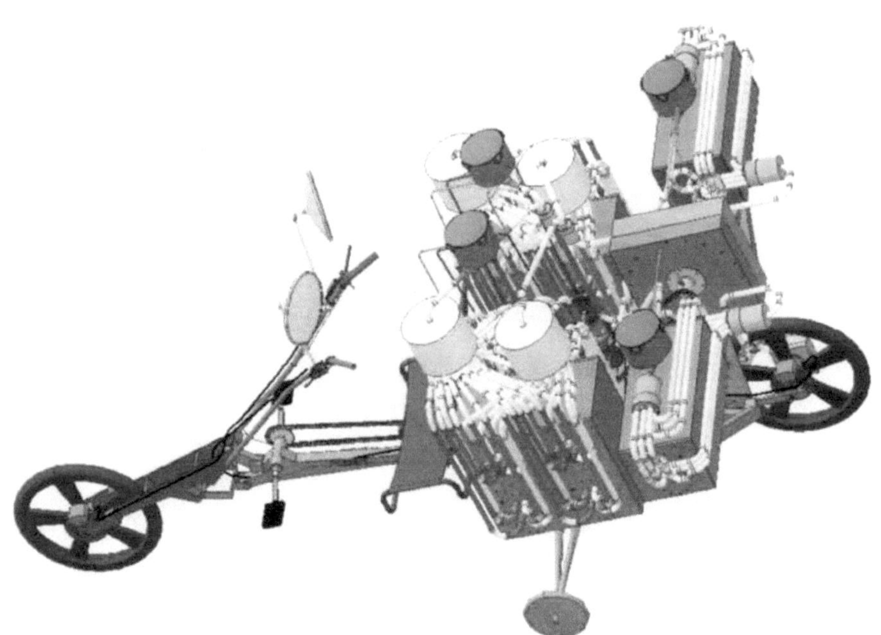

(In this case $A = 2$.)

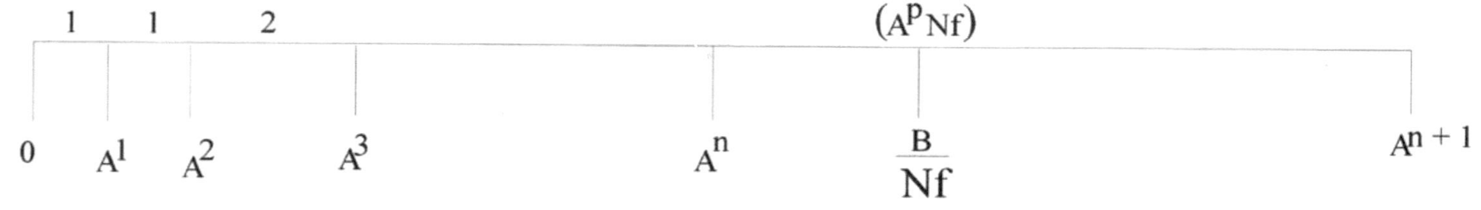

1 1 2 $(A^p Nf)$

$0 \quad A^1 \quad A^2 \quad A^3 \quad A^n \quad \frac{B}{Nf} \quad A^{n+1}$

$A^p Nf \neq \frac{B}{Nf}$ when $Nf \neq 1$.

$$Nf\,A\wedge\left(\frac{\frac{B}{Nf} - A^n}{A^{n+1} - A^n}\right) + n = A^p Nf = B$$

Solve for p.

n = an integer whole number.

N = density

f = volume

B = Force, lbs.

A = base, 2.

The rational solution to making the product, meshing gear teeth volume a minimum quantity is to make the clearance first half of the calculated clearance which will reduce the product, meshing gear teeth volume by half.

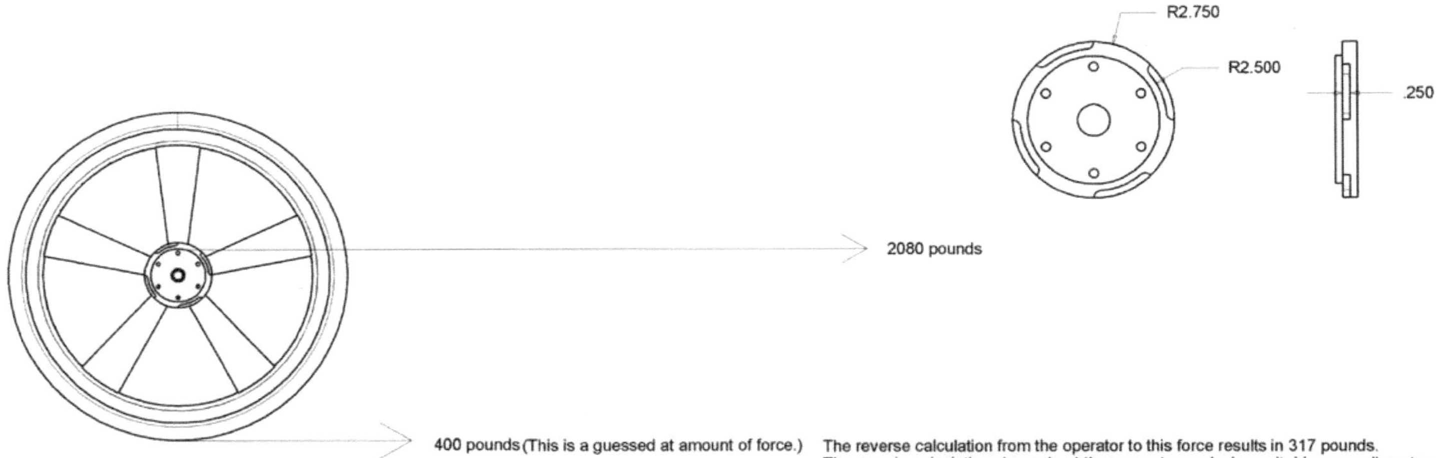

R2.750

R2.500

.250

2080 pounds

400 pounds (This is a guessed at amount of force.)

The reverse calculation from the operator to this force results in 317 pounds.
The error in calculations is made at the power transmission unit drive gear diameter.

$2.75^2 \times pi \times .25 - 2.5^2 \times pi \times .25 = 1.031$ cubic inches.

the density of hydraulic fluid is: 53.9 pounds per cubic foot.

$53.9 / 12^3$ cubic inches per cubic foot = .031 pounds per cubic inch.

$$\frac{\frac{2080 \text{ pounds}}{.031 \times 1.031} - 2^n}{2^n} + n = p = 15.98606$$

1.031 x 15.98606 x 2 (double the count of blade vanes) x 2 (only half the impeller is used for performance) + impeller clearance volume x p = initial meshing gear teeth volume for one circuit of revolution of the drive gear (divided by three driven gears) / 16 revolutions per two cranks.

Initial meshing gear teeth volume / 3 / the working depth area of the driven gear meshing gear teeth / 2 power transmission units = the gears depth.

The impeller to the secondary Power transmission unit also applies the same formulas.
There are no denominators except 3 meshing gear teeth and two (2) power transmissions.
The crank revolution ratio is 2:1 so the chainring may be 1/2 the diameter in the drawings.
in this case the force at the crank is acceptable at 150 pounds, which will allow the road/tire contact force to be 400 pounds with pedal force to spare..

(In this case A = 2.)

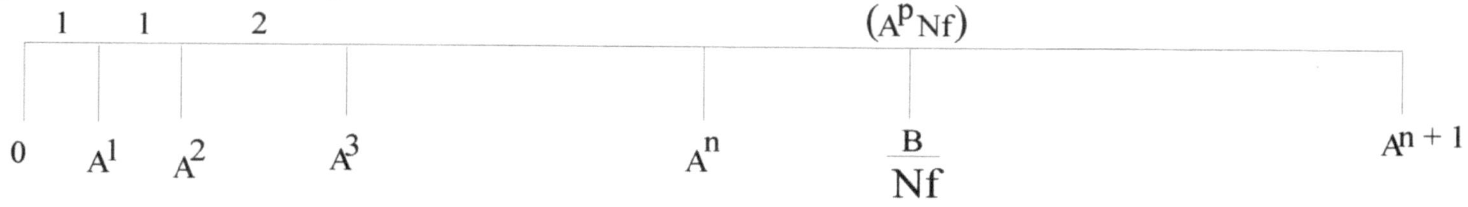

1 1 2 $(A^p Nf)$

$0 \quad A^1 \quad A^2 \quad A^3 \quad A^n \quad \frac{B}{Nf} \quad A^{n+1}$

$A^p Nf \neq \frac{B}{Nf}$ when $Nf \neq 1$.

$$Nf\, A \wedge \left(\frac{\frac{B}{Nf} - A^n}{A^{n+1} - A^n}\right) + n = A^p Nf = B$$

Solve for p.

n = an integer whole number.

N = density

f = volume

B = Force, lbs.

A = base, 2.

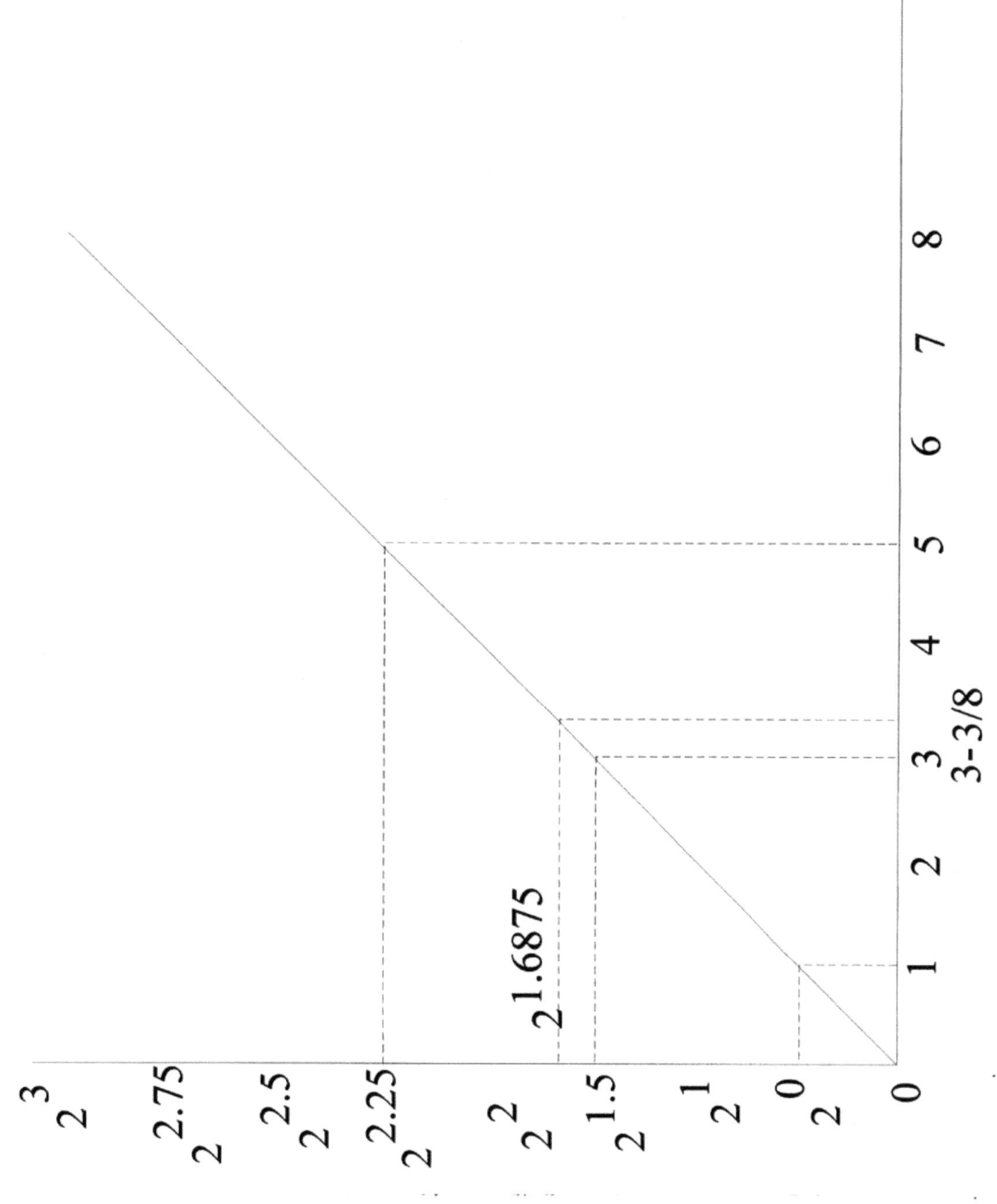

2^{3}
$2^{2.75}$
$2^{2.5}$
$2^{2.25}$
2^{2}
$2^{1.6875}$
$2^{1.5}$
2^{1}
2^{0}
0
1
2
3
3-3/8
4
5
6
7
8

$$\frac{\text{Bird's Inertia}}{\left(\dfrac{\dfrac{\text{air density x displacement of bird}}{\text{swept volume at NSL}}}{\text{one displacement volume}}\right)} \text{ x bird's inertia} = \text{impact reaction force of bird @ NSL}$$

$$\frac{\text{displacement distance of bird @ NSL}}{\dfrac{\text{inertia of displacement of bird at air density @ NSL}}{\text{inertia of displacement of bird x air density at rest}}} = \text{one unit displacement for one unit volume of the bird}$$

$$\text{Bird's inertia at rest x } \frac{\text{inertia of displacement of bird at air density @ NSL}}{\text{inertia of displacement of bird x air density at rest}}$$

$$= \text{impact reaction force of material bird @ NSL}$$

$$\frac{\text{Impact reaction force @ NSL}}{\left(\dfrac{\text{inertia of displacement of bird at air density @ NSL}}{\text{inertia of displacement of bird x air density at rest}}\right)} = \text{impact reaction force of material bird at 1 unit displacement distance, should equal 1.}$$

$$\frac{\text{Time the transit of units of displacement volume @ NSL}}{\left(\dfrac{\text{inertia of displacement of bird at air density @ NSL}}{\text{inertia of displacement of bird x air density at rest}}\right)} = \text{rate at 1 unit displacement distance on take off}$$

Increments of rate between different species of bird is not equal, on the ground or at any altitude.

The bird displaces the swept volume divided by * which is equal to 1 unit coefficient cubit of one bird inertia.

* = the denominator in the above equation.
NSL = normal speed @ lift.
The initial moment of applied equal and opposite reaction force to the bird displacing over the coefficient fraction of distance is equal to 1 unit cubit equal to one unit of birds inertia. The bird inertia (moment of equal and opposite reaction force at NSL is equal to the inertia of the bird x *.

The bird initially starts out at 0 displacement inertia, goes to 1, goes to 1×2^p. * may be referentially p. The displacement of the bird also starts out at 1 inertia and goes to 1×2^p.

Impact reaction force of bird @ 1 unit displacement should equal one volume of the bird's displacement in inertia @ air density.

One swept volume of the bird sustains NSL.

$$\frac{\text{One swept volume}}{*} = \text{swept volume at rate on take-off, start from 0 rate.}$$

2^p is corelevant with respect to the denominator:

$$\left(\frac{\dfrac{\text{inertia of displacement of bird at air density @ NSL}}{\text{inertia of displacement of bird x air density at rest}} - 2^n}{2^n} \right) + n = 2^p$$

This is an exercise. Should it prove true then more power may be applied.

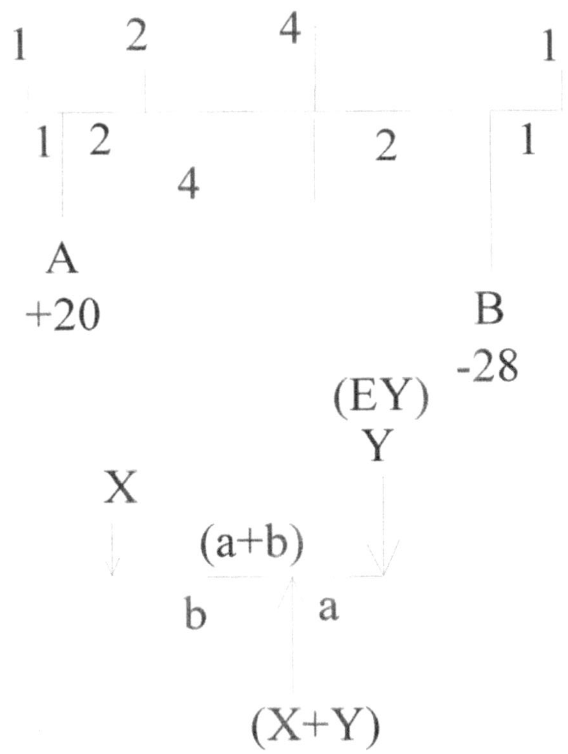

1 2 4 1

1 2 4 2 1

A
+20

B
-28

(EY)
Y

X

(a+b)

b a

(X+Y)

A=-1(1)+2(4)+6(2)+9(1)

B=+1(1)-2(2)-6(4)-9(1)

Find the fulcrum two loads at a time eliminating one by one, and then the two distances remaining from the fulcrum to A & B are the percentages of the sum of the loads at A & B.

Y= 2, 4, other 1, etc. = sum of Y = EY
X= first 1.

$$\frac{X(a+b)}{(X+Y)} = a$$

$$\frac{a}{(a+b)}(X+EY) = A \qquad \frac{b}{(a+b)}(X+EY) = B$$

Using the same formula from the previous page, calculate the fulcrum(s):

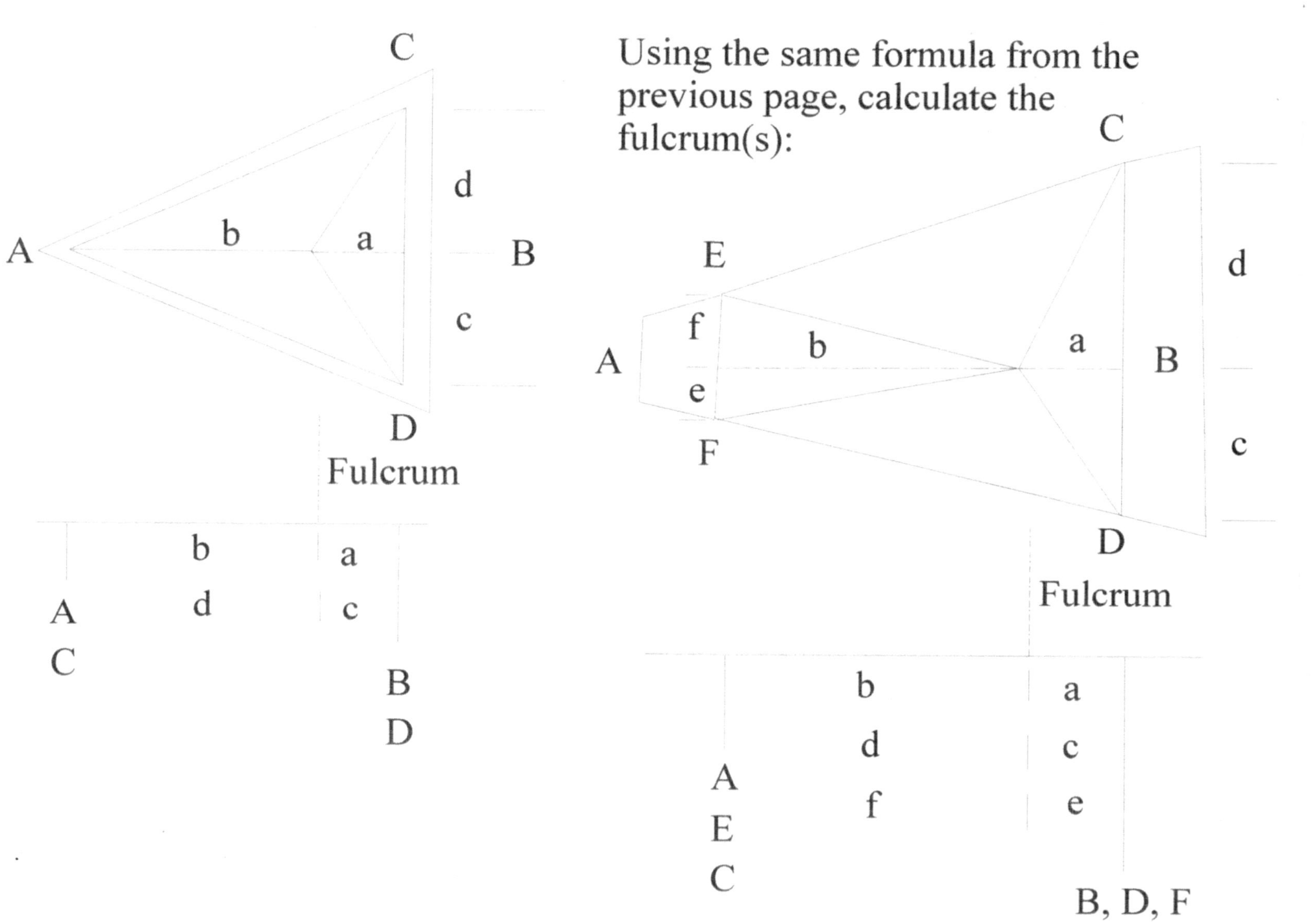

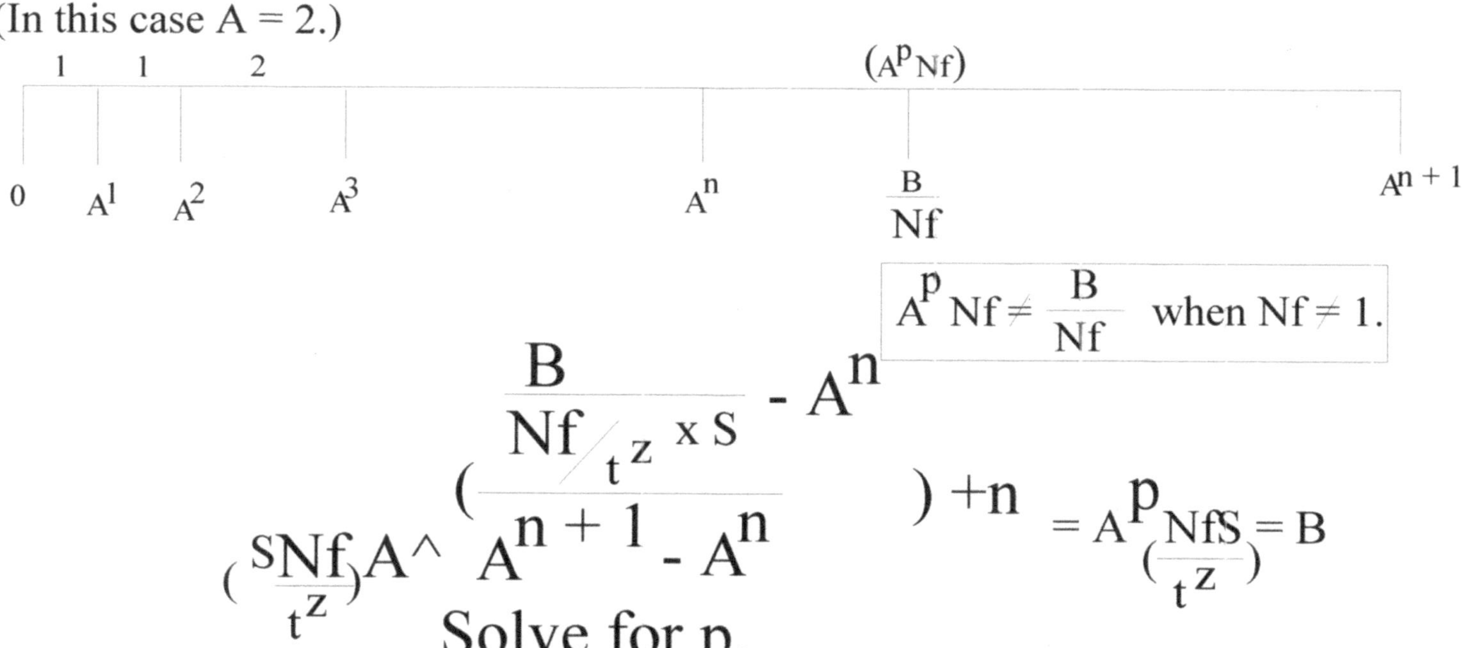

$$\left(\frac{SNf}{t^z}\right)A\wedge\left(\frac{\frac{B}{Nf/t^z \times S} - A^n}{A^{n+1} - A^n}\right) + n = A^{p}\left(\frac{NfS}{t^z}\right) = B$$

n = an integer whole number.

N = density in lbs./cu.ft.

f = volume in cu. in.

B = Force in ounces

S = oz./lb. in base 2

A = base, 2.

$t^z = 12^3$ cubic inches/cubic foot in base two.

S = some unknown quantity in any other base.

Once again the exponent power function of the calculator is causing an error.

12^3 is used in base 2. What other units of measure would be used for other bases?

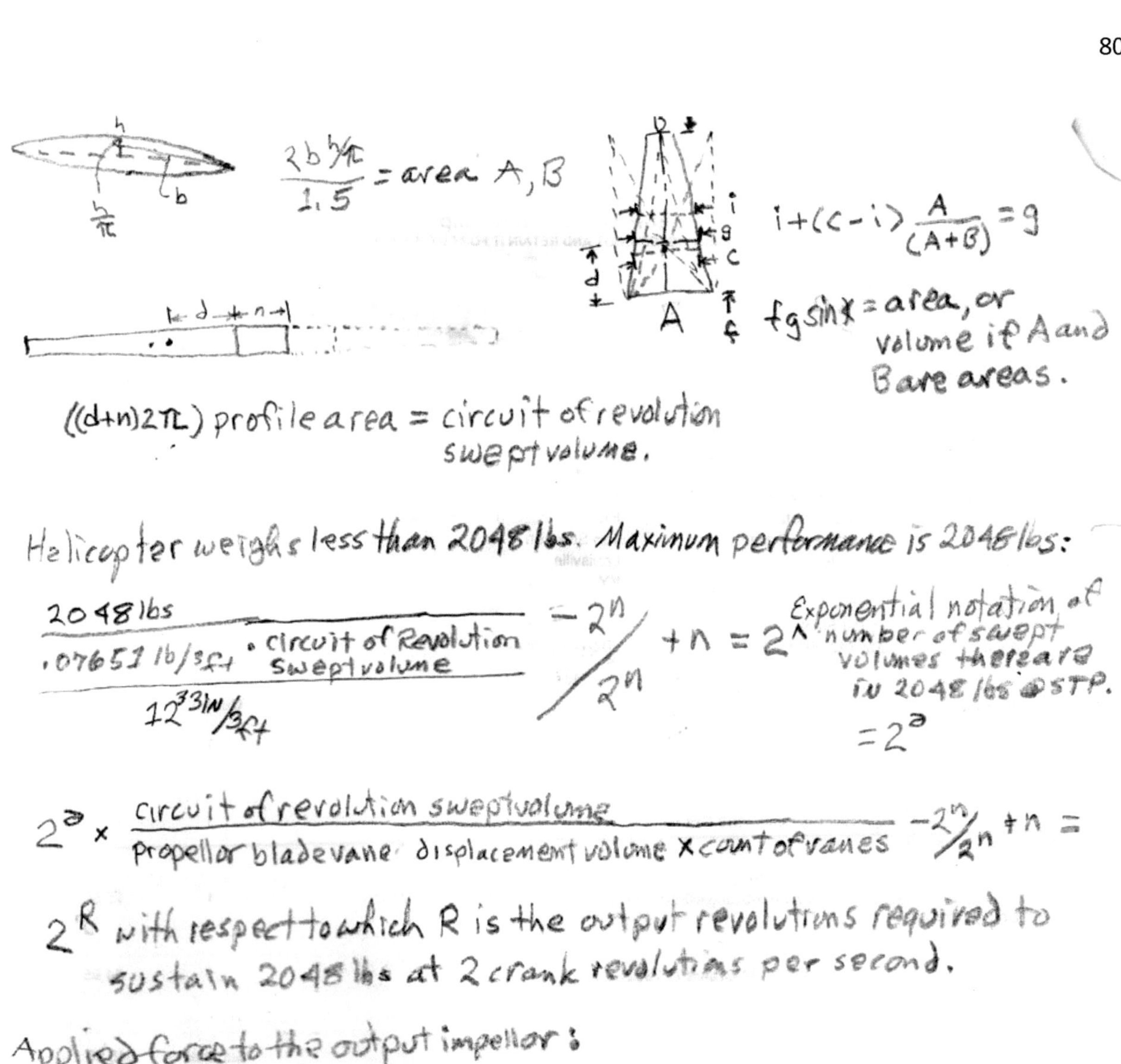

((d+n)2π) profile area = circuit of revolution swept volume.

Helicopter weighs less than 2048 lbs. Maximum performance is 2048 lbs:

$$\frac{2048\text{ lbs}}{\frac{.07651\text{ lb/}^3\text{ft}\cdot\text{circuit of Revolution swept volume}}{12^3\ {}^3\text{in}/{}^3\text{ft}}}-2^n\Big/2^n+n=2^{\wedge}$$

Exponential notation of number of swept volumes there are in 2048 lbs @STP. $=2^a$

$$2^a\times\frac{\text{circuit of revolution swept volume}}{\text{propellor blade vane displacement volume}\times\text{count of vanes}}-2^n/2^n+n=$$

2^R with respect to which R is the output revolutions required to sustain 2048 lbs at 2 crank revolutions per second.

Applied force to the output impellor:

$$\frac{d+\text{the lift blade hub radius}}{\text{hub radius of the output impellor}}\times(2048+2048(\sin\text{ airfoil plane angle})$$

= applied force to the output impellor blade vanes hydraulic fluid. at maximum performance.

The applied impellor vanes volume to the applied maximum force:

(impeller vanes maximum radius² · π · depth − impellor hub radius · π · vane depth) · 2 ·

$$\frac{\text{applied maximum performance}}{\frac{(\text{in parenthesis above})\cdot 69.5\text{ lb/}^3\text{ft}}{12^3\ {}^3\text{in}/{}^3\text{ft}}}-2^n/2^n+n\cdot R+(\text{area of passageway around impellor vanes}\cdot\pi\cdot\text{depth})$$

= flow volume in ³in. to the lift impellor.

$\frac{\text{[illegible] in/16 rev}}{\text{PTU center gear mgT area} \times 6}$ = 5.08in gears depth.

$\frac{\text{The flow volume in } ^3\text{in to the lift impellor/16 rev. of the secondary PTU}}{\text{PTU center gear mgt area} \times 6 \text{ mgt}}$

= gears depth.

The secondary PTU impellor radius begins calculations arbitrarily:

$\frac{\text{applied force to lift impellor at maximum performance} \times \text{secondary PTU center gear radius}}{\text{secondary PTU impellor hub radius}}$

= maximum applied force to secondary PTU impellor blade vanes.

The gear arrangement is 7" radius at the primary PTU, 3½ radius at the crank, and the crank turns a 1¾" radius

So; 12,250 lb × $\frac{1.4''}{12''}$ × 1.4"/7" × 7"/1.75"/2

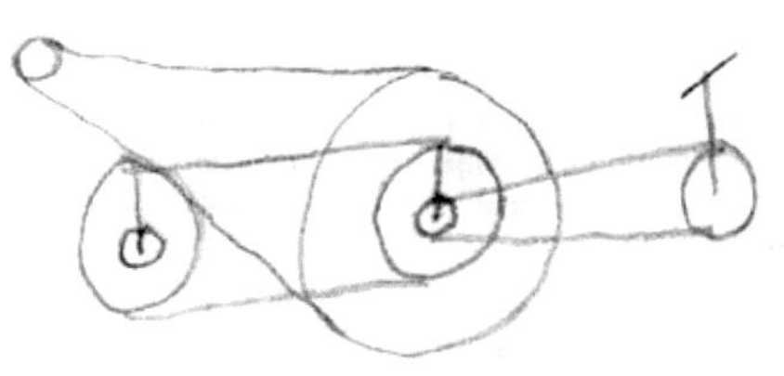

= operator's force at the crank.
÷ 200 lbs desired operator's force
× 12" impellor radius = 34.3" impellor radius resulting in 200 pounds at the operator's force at the crank.

34.3
34.3 + .25 = 34.55
34.6125, 34.55, .25
34.6125, 34.3, .1875

~~Calculate the radius of the secondary PTU impellor blade vanes volume = the same impellor hub radius volume at the blade vanes. Add the clearance passageway volume, multiply by 16 divide by 6 mgt areas, divide by 2~~ ~~count of primary ptus~~ → ~~PTUs, divide by mgt area of the center gear of the primary PTU = primary PTU gears depth.~~

Calculate the secondary PTU impellor bladevanes volume, multiply by the count of revolutions, multiply by (the applied force to the impellor divided by the quantity the impellor blade vanes volume times the hydraulic fluid density. Subtract 2ⁿ. Divide by 2ⁿ. Add n). Multiply that number. Calculate the clearance volume of the same impellor, multiply by the same base 2 quantity exponent, add to previous product, divide by primary PTU centergear mgT area, divide by 6, divide by count of PTUs = gears depth.

$$\frac{\text{Lift blade impellor applied load}}{\text{hydraulic fluid density} \times (\text{impellor gallery volume} - \text{impellor volume})} \quad -2^n/2^n + n$$

$12^3\ 3in/3ft$

$= V$

Impellor blade vanes volume $\%R \cdot V \cdot R$ + passageway clearance volume

~~$\times V \times 1/R$~~ $\times R$ + passageway clearance volume $\cdot V \cdot 1/R$ = initial mgTV.

In the case of the secondary PTU, R = 16 revolutions per cranks^2 per second.

$$\frac{\text{1273.15 }^3\text{IN/16 Rev}}{\text{PTU center gear mgT area} \times 6} = \text{5.08 in gears depth.}$$

$$\frac{\text{The flow volume in }^3\text{in to the lift impellor/16 rev. of the secondary PTU}}{\text{PTU center gear mgt area} \times \text{6 mgt}}$$

= gears depth.

The secondary PTU impellor radius begins calculations arbitrarily:

$$\frac{\text{applied force to lift impellor at maximum performance} \times \text{secondary PTU centergear radius}}{\text{secondary PTU impellor hub radius}}$$

= maximum applied force to secondary PTU impellor blade vanes.

The gear arrangement is 7" radius at the primary PTU, 3½ radius at the crank, and the crank turns a 1¾" radius

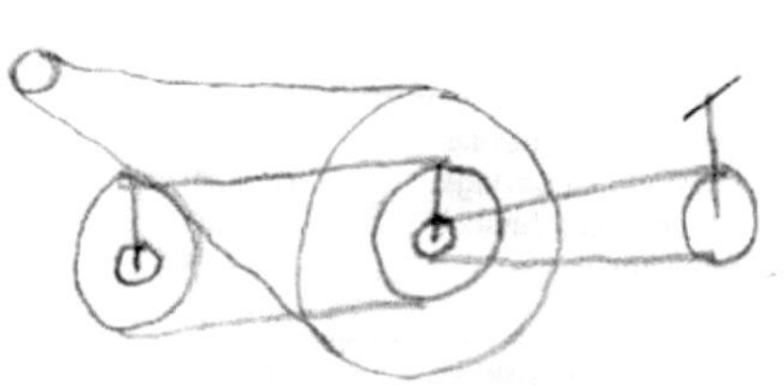

So; 12,250 lb × 1.4"/12" × 1.4"/7" × 7"/1.75"/2

= operator's force at the crank.

÷ 200 lbs desired operator's force

× 12" impellor radius = 34.3" impellor radius resulting in 200 pounds at the operator's force at the crank.

34.3
34.3 + .25 = 34.55
34.6125, 34.55, .25
34.6125, 34.3, .1875

~~Calculate the radius of the secondary PTU impellor blade vanes volume = the same impellor hub radius volume at the blade vanes. Add the clearance passageway volume, multiply by 16 divide by 6 mgt areas, divide by 2~~ (count of primary ptus →) ~~PTUs, divide by mgt area of the center gear of the primary PTU = primary PTU gears depth.~~

Calculate the secondary PTU impellor bladevanes volume, multiply by the count of revolutions, multiply by (the applied force to the impellor divided by the quantity the impellor blade vanes volume times the hydraulic fluid density. Subtract 2n. Divide by 2n. Add n). Multiply that number. Calculate the clearance volume of the same impellor,

$34.55^2\pi \cdot .25 - 34.3^2\pi \cdot .25 = 13.51867\ in^3$

$34.6125^2\pi \cdot .1875 - 34.3^2\pi \cdot .1875 = 12.68525\ in^3$ } +

$34.6125^2\pi \cdot .25 - 34.55^2\pi \cdot .25 = 3.39501\ in^3$

$$\frac{500}{\frac{69.5 \cdot 13.51867}{12^3}} - 2^n/2^n + n = 9.79608$$

$13.51867 \cdot 9.79608 \cdot 16 \times 2 = 4237.75794\ in^3$

$16.08026 \cdot 9.79608 + \frac{16.08026 \cdot 9.79608}{16\ rev.} = 167.36873$

$$\frac{4405.12667\ in^3}{2.61096\ in^2 \cdot 4\ rev \cdot 3_{mgTUs} \cdot 6_{PTUs}} = 23.43288''\ \text{initial depth}$$

Clearance volume:

$(.4035^2\pi - .14^2\pi) \cdot 23.43288\ in \times 6 = 4.33406\ in^3$ } +

$(1.8535^2\pi - 1.85^2\pi) \cdot 23.43288\ in \cdot 3 \cdot 6 = 17.17625\ in^3$

$2.61096 \times 23.43288 \times 3 \times 6 = 1101.28162$

$$\frac{1101.28162\ in^3}{1101.28162 - 21.51031} \times 1101.28162 = 1123.22044\ in^3$$

(1.01992)

$23.43288 \times 1.01992 = 23.89966''$ primary PTU gears depth.

$\frac{2327\ ^3in}{7.83287\ ^2in}$ = (18.6 in gears depth.)

Tail rotor primary PTU

4 Rev/2 cranks/sec

4 | 2 | PTUs

2

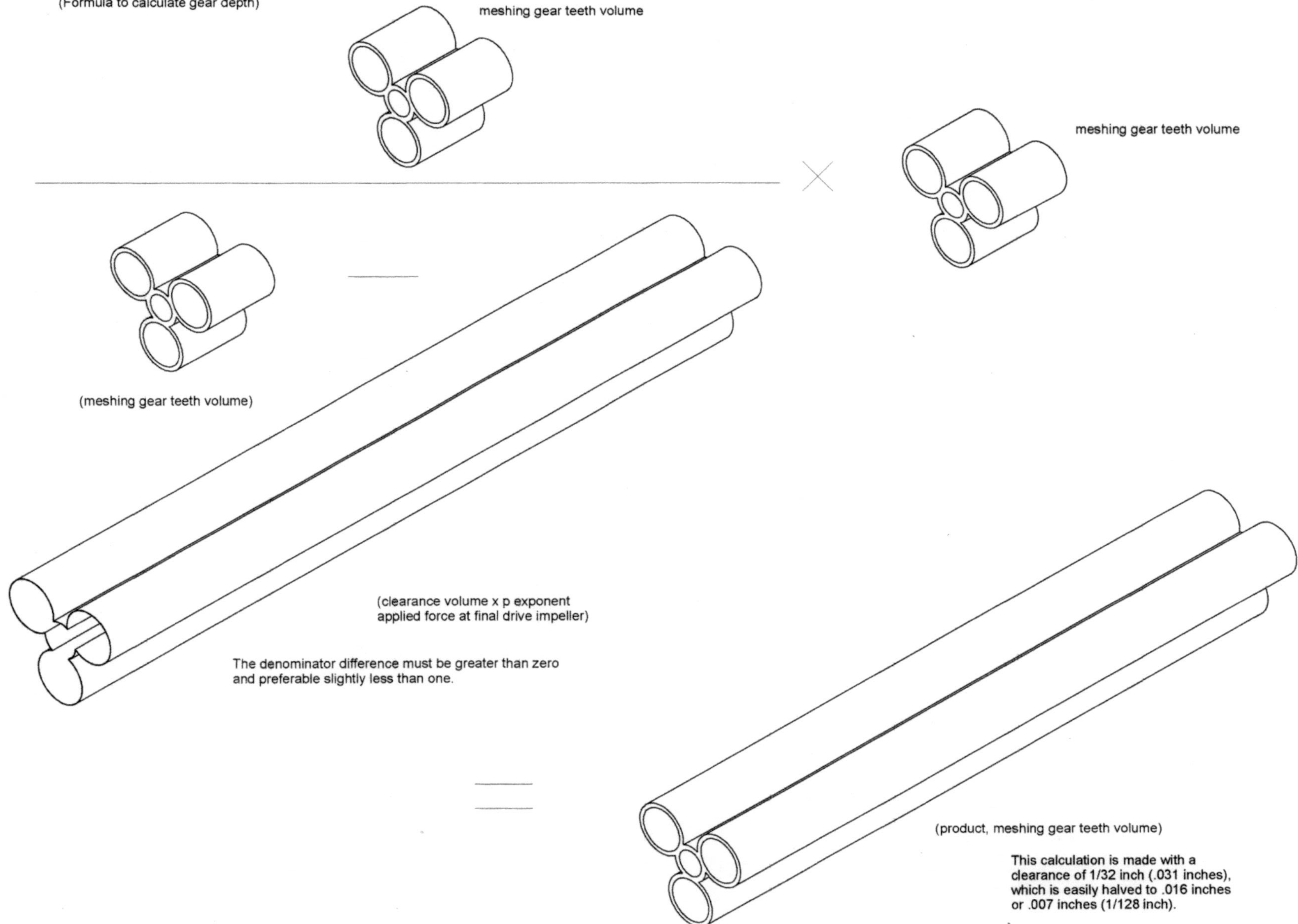
(Formula to calculate gear depth)
meshing gear teeth volume
meshing gear teeth volume
(meshing gear teeth volume)
(clearance volume x p exponent
applied force at final drive impeller)
The denominator difference must be greater than zero
and preferable slightly less than one.
(product, meshing gear teeth volume)
This calculation is made with a
clearance of 1/32 inch (.031 inches),
which is easily halved to .016 inches
or .007 inches (1/128 inch).

$$\frac{\text{Applied pounds}}{\text{metal shear strength}} = {}^{2}\text{in}$$

$$2 \times \sqrt{\frac{{}^{2}\text{in} + \text{pi}\, r_1^2}{\text{pi}}} = r^2 \times 2 = \text{diameter}_{r_2}$$

$$2 \times r_1 = \sqrt{\frac{r_2^2 \times \text{pi} - {}^{2}\text{inches}}{\text{pi}}} \times 2 = \text{diameter}_{r_1}$$

(Formulas For Tubing Wall Thickness)

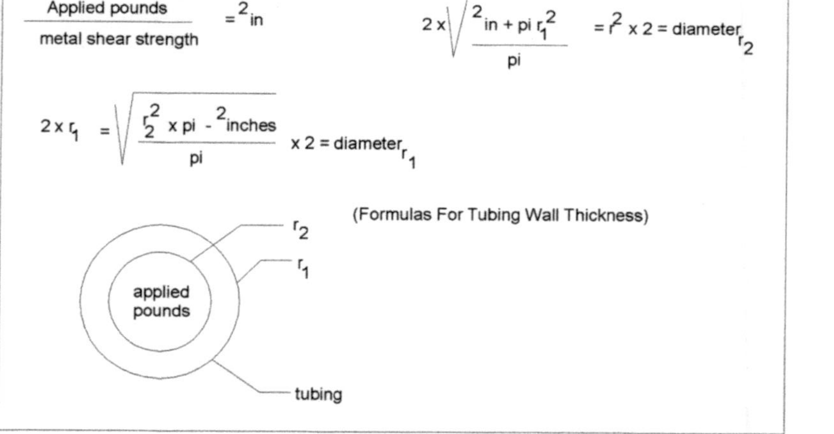

X : N : : X : N

300 : 48 : : 150 : 24

X : N : : 150 : 24

N = number of teeth in sprocket

N approx. = 36

X = resulting force closest to approximately 220 pounds.
Force varies per occupant.

N / 24 x the circumference of the 24 tooth chainring = the N tooth chainring circumference

9.9375 in. diameter x pi = 31.22 in. circumference.

2.625^2 x pi x .25 - 2.25^2 x .pi x .25 = .50315 cubic inches x 2 double the blade vanes volume = 1.00629 cubic inches.

$$\frac{\dfrac{2080 \text{ lbs.}}{\dfrac{63.9 \text{ lbs. / cu. ft.}}{12^3 \text{ cu. in. / cu. ft.}} \times 1.00629 \text{ cu. in.}}}{2^n} - 2^n + n = 15.72735 \qquad n = 15$$

(Clearance volume x 15.72735 + 1.00629 x 15.72735) x 60.011649 revolutions per second/ 16 revolutions of the power transmission unit = the secondary meshing gears volume.

The same coefficients apply to the center "box" with respect to the initial meshing gears teeth volume while the center "box" only has 16 revolutions. Torques are readily calculable.

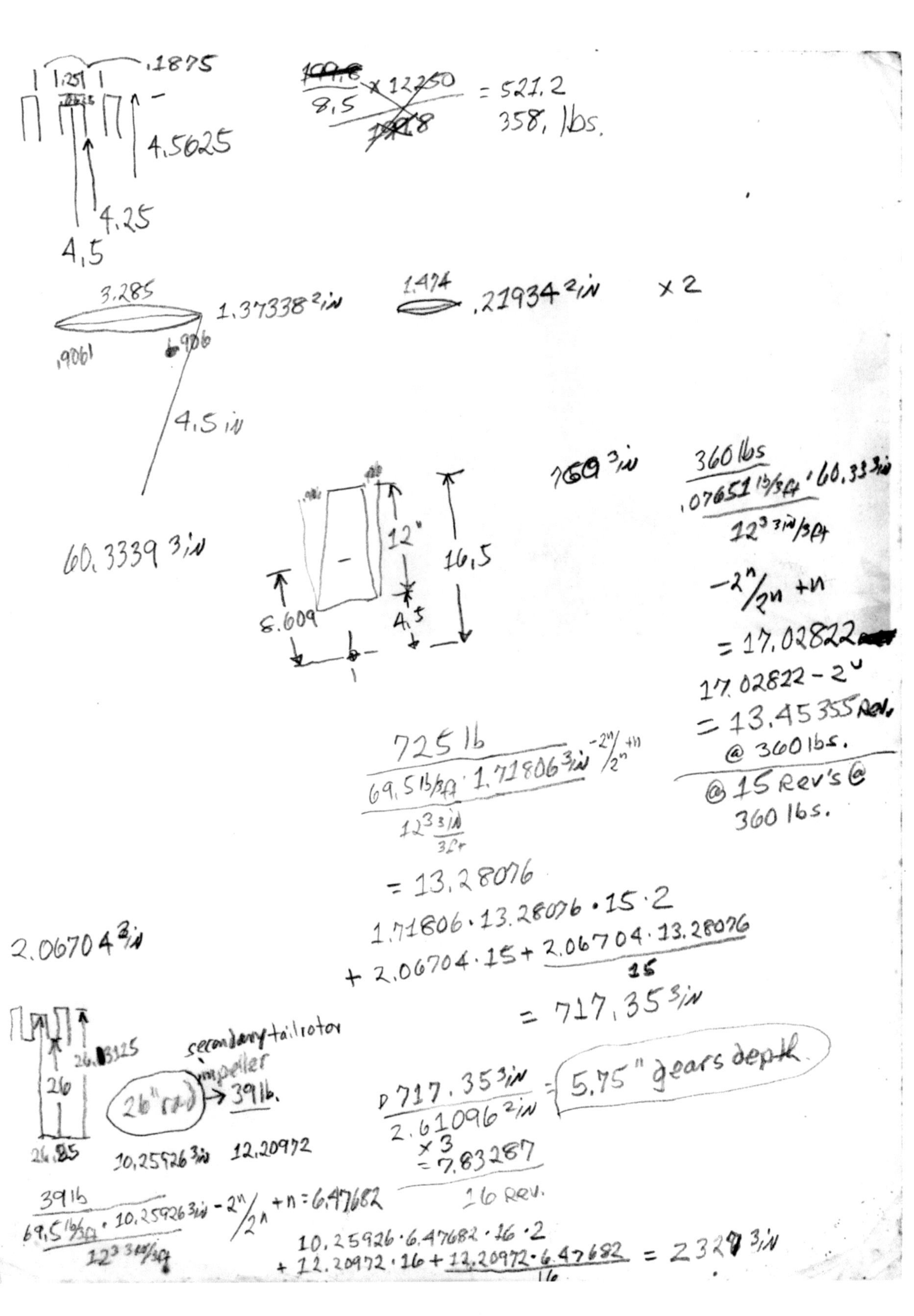
.1875
4.5625
4.25
4.5
= 521.2
358. lbs.
3.285
1.37338 ²in
.21934 ²in
x 2
4.5 in
60.3339 ³in
360 lbs
12"
16.5
8.609
= 17.02822
17.02822 - 2
= 13.45355 Rev.
@ 360 lbs.
@ 15 Rev's @
360 lbs.
725 lb
69.5 lb/ft³ · 1.71806 ³in
= 13.28076
1.71806 · 13.28076 · 15 · 2
+ 2.06704 · 15 + 2.06704 · 13.28076
2.06704 ³in
= 717.35 ³in
secondary tailrotor impeller
26" rad
391b.
717.35 ³in
2.61096 ²in
x 3
= 7.83287
16 Rev.
5.75" gears depth
10.25926 ³in
12.20972
391b
= 6.47682
10.25926 · 6.47682 · 16 · 2
+ 12.20972 · 16 + 12.20972 · 6.47682

hydraulic bearing fluid density is in pounds per cubic foot.
clearance volume is in cubic inches.

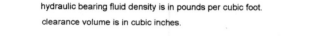

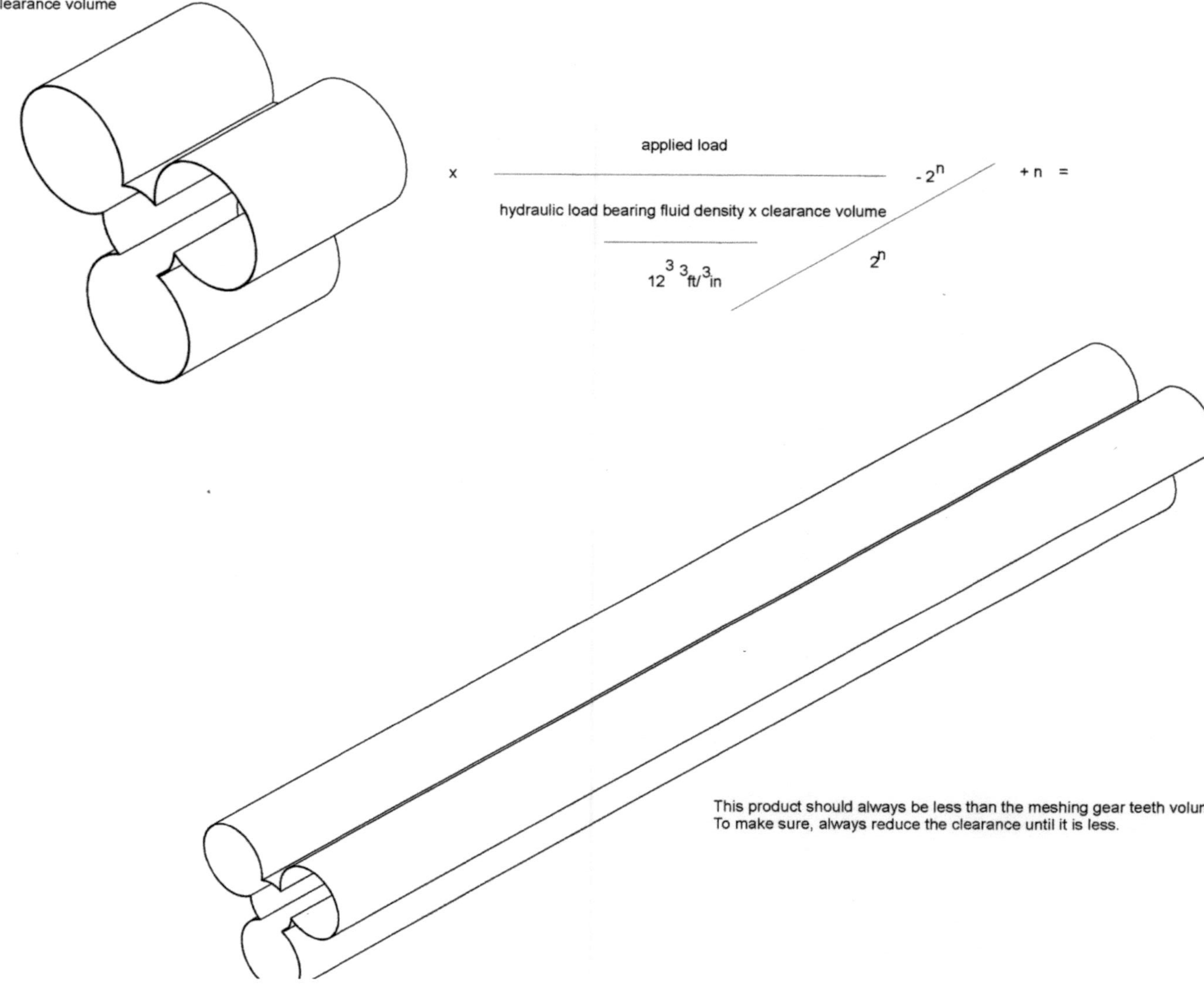

$$\times \frac{\dfrac{\text{applied load}}{\dfrac{\text{hydraulic load bearing fluid density} \times \text{clearance volume}}{12^3\ {}^3\text{ft}/{}^3\text{in}}} - 2^n}{2^n} + n =$$

This product should always be less than the meshing gear teeth volume.
To make sure, always reduce the clearance until it is less.

This is the formula for finding the center of an oblique trapezoid, eg: a wing.

However, it has since been discovered that wings, e.g. the delta wing, is balanced by the aA/b = B method although the center of the delta wing volume geometry is different from its aA/b = B balance equilibrium moment. Swept back delta wings are still balanced by the aA/b = B formula although for symmetry it is easier to make the plane fly straight if it looks the same on the other side and therefore even if the wings are swept back and the delta seems more difficult to figure out, finding the aA/b = B equilibrium moment for the geometry of the wingS will put the aA/b = B cantilever in the center of the airplane, and where the formula balances is where the delta wings geometry bisector plane orbits with the wind. If the airplane (vessel, vehicle, etc.) has multiple delta airfoils all their individual aA/b = B fulcrums are individually found and then all together their Em = 0 is where the delta air foil lift is balanced. Including the remaining geometry in the aircraft body volume and all of the ups and downs forces throughout the geometric displacement of the model and including the delta Em = 0, again find the remaining Em = 0 and continue to further the resolution of details involving balancing the model.

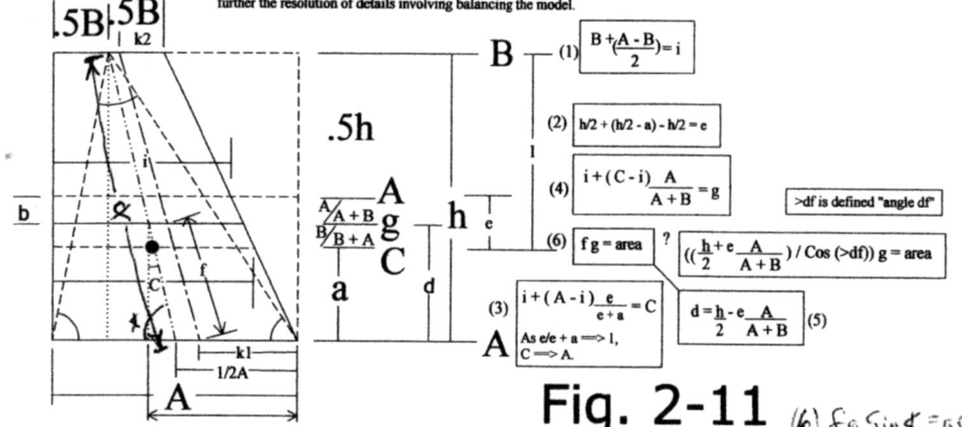

Fig. 2-11

(6) fg Sin∡ = area

Finding the center(s) for the rib area(s) applies the rib area analysis for the airfoil plane angle and its opposite centers and their line congruent to WR-parallel and referring the centers' dimensions back to the trapezoid drawing.

$$\left[\ \frac{B}{A}\left(A - \left(\frac{h}{h} - \frac{a}{h}\right)A = \frac{B}{A}(A - C) = \text{the difference of } g - C.\ \right]\right.$$

$$\left[\ \frac{B}{A}\left(A - \left(\frac{h}{h} - \frac{a}{h}\right)A\right) + \left(\left(\frac{h}{h} - \frac{a}{h}\right)A = \frac{B}{A}(A - C) + C = g.\ \right]\right.$$

$$[\ gh = \text{volume when A, B, C, and g are areas.}\ ]$$

$$\left[\ (h - .5h - a)\frac{C}{A} = b\ \right] \qquad [\ .5h - b = d\ ]$$

$$\left[\ \left(\frac{h}{h} - \frac{a}{h}\right)A = C\ \right]$$

Original works left for speculation

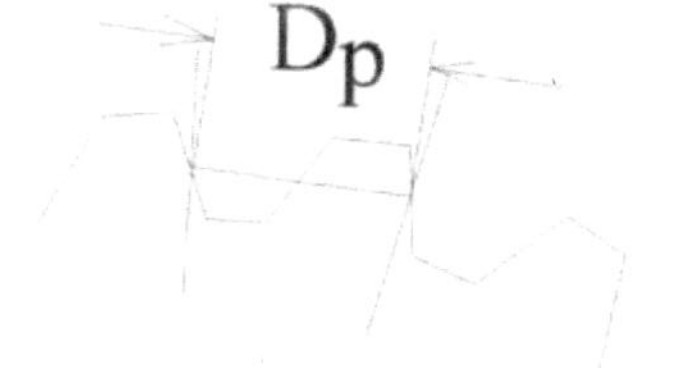

Whole depth of tooth

D_p = 1/8, 1/4, 3/8, etc.

This is American Standard Version.

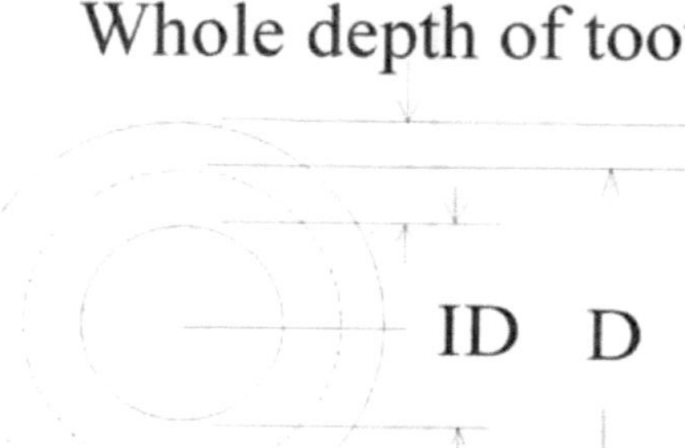

2 x pi = 6.2832

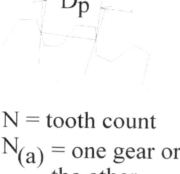

N = tooth count

$N_{(a)}$ = one gear or the other

D = basic diameter

D_p x .6866 = Whole depth of tooth

$$\frac{(N_1 + N_2) \times D_p}{6.2832} = \text{Centers Distance}$$

O.D.$_1$

O.D.$_2$

$$\frac{D_p N}{pi} = D$$

$$\frac{(N + 2) \times D_p}{pi} = O.D.$$

$$\frac{\left(\frac{(N + 2) \times D_p}{pi} - D_p \times .6866\right) \times 2}{2} = I.D.$$

$$.6865 = \frac{703}{1024}$$

$$\frac{704}{1024} = \frac{11}{16}$$

.6865 may be used to avoid copyright infringment.

Add + 0.014 inches to O.D. for clearance.

...I don't know where .6866 came from.

First HPH withou CAD and calculations or whatever, was destroyed. It is on CD but it is encrypted. All my model planes are destroyed. Everything is at the Library of Congress under Fluid Mechanics of Inertia all inclusive supplements thereto 1 through 7 (first through seventh). This will be the determination to continue the all inclusive Fluid Mechanics of Inertia:

The component variable of an airplane's displacement volume of air at the density at altitude multiplied by its volume is equal to one unit cubit. The displacement of the blade vanes volume of air at altitude times the air density at atltitude equals one unit cubit. The same is true if the vanes are displacing fire, the density of the fire times the displacement of the volume of the blade vanes is equal to one unit cubit. The applied fold force of the combustion of the fire is the numerator in the equation. The applied force of the blade vanes at zero elapsed time is the numerator with respect to the cubit of blade vanes volume and air density at altitude. The pitch of the blade vanes displaces a particular quantity of volume at limits with respect to the required volume to be equal to the air density times the blade vanes volume times the fold proportion required to equal the weight of the airplane plus the drag but the blade vanes have to have pitch because without drag the vanes will not develop thrust. It is amazing that a such a small powerplant can develop so much power that the blade vanes of an airplane or helicopter can sustain lift at no elapsed time while applying elastic forces to the blade vanes volume at air density by only applying pitch to the vanes and it is because of avgas that the engines do not fowl out during sustained flight because the engines must sustain this applied force throughout the enitire life of the aircraft in flight which is like driving an automobile vertically at a constant speed. Converting these limits of tolerances over to human powered flight and the power transmission unit assembly to operator's force results in the impending human powered heilicopter and super slip-stream bicycle and propetual motion engine. All relavent applications of base two mechanics include every feasible means of locomotion in to which a person may enter and operate said means of locomotion. The applied force inside and operating aircraft engine is sustained at no elapsed time and the combustion performance force applied to the power train lever delivers the required coeffiecient fold force numerator to the cubit denominator deriving the fold coefficient and when multiplied to the cubit equals the required force of the airplanes weight plus drag at the required altitude for air density. All of this happens at no elapsed time. The applied force in the engine results at no elapsed time. Thus this is a considerable force and can be calculated in reverse from the required load at the blade vanes to the internal combustion force in zero elapsed time. Due to the fact that the applied force at the blade vanes and the lever arm torque in the power train to the internal combustion increases at the internal cumbustion, the internal combustion force is greater than the applied force at the blade vanes, and again the powerplant is more powerful than the required force needed for it's size. Amazing how powerful the internal combustion engine is considering how relatively light it has to be. Jet engines are even more confusing. How do you get power out of something that can't even hang on to its combustion? Rocket engines are even worse. Making a fire turn a fan most efficiently I will leave to the experts, I only express that limits of tolerances be applied with respect to any free volume that may be escaping between the vanes and is not accounted for as applied load to the blade vanes volume.

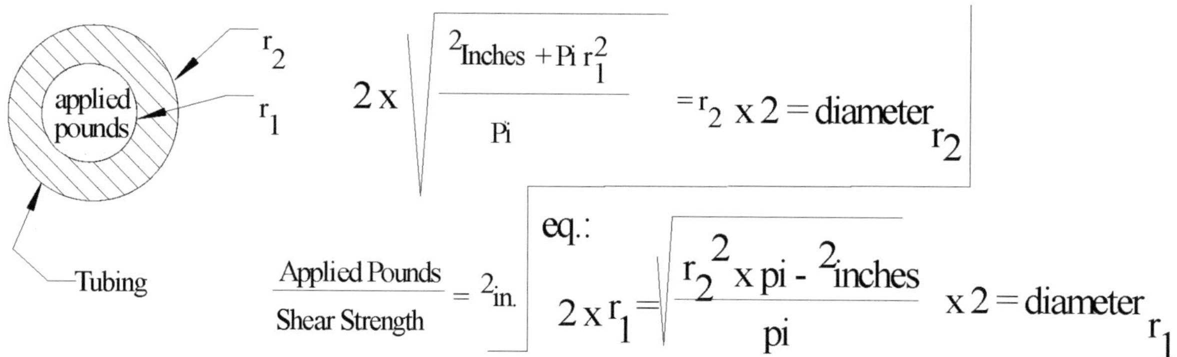

For a helicopter: If the pitch angle of the blade only needs to apply 1/p to sustain lift and the vector applied congruent to the direction of the blades at normal speed at left equal to 1024 pounds is equal to p and the pitch angle of the blades is then p Cos>2 Cos >1 = 1.

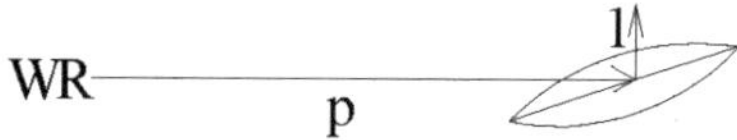

This configuration should sustain hover of a 1024 pound helicopter.
This is a speculation and is not proven. The ratio may need to be reduced to greater than 1/p. e.g. 1/8 or 1/6 etc.

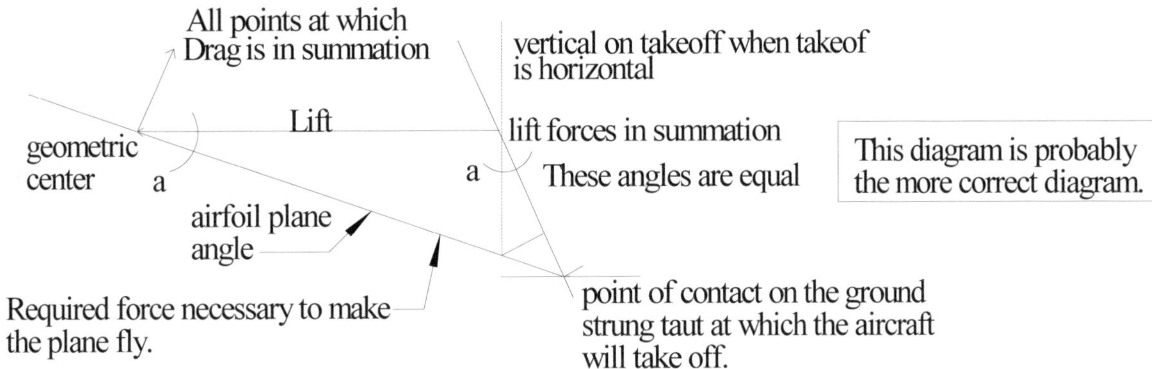

This is perhaps the minimum at which the drag will be sufficient to lift the aircraft provided the aircraft is designed to have the drag to weight ratio to be conformed to this sine ratio.

Notice this strung taut point may be applied even if the aircraft is below ground since the point will be strung taut and thus the aircraft will climb, as if it were on a string. A more accurate kind of "string" is a rubber band.

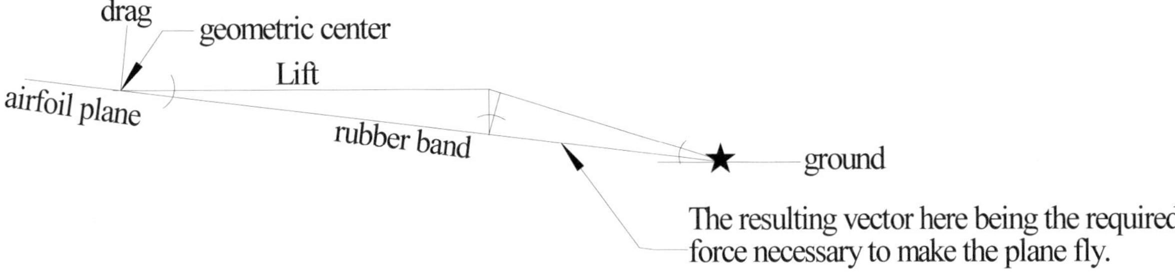

The previous page seems to be bogus. It seems that any remaining efforts put in to evaluating Fluid Mechanics of Inertia are futile. I seem to have everything down.

In reference to the wing and the amount of drag in proportion to the lift needed, the airfoil plane angle will reduce the swept volume required to sustain lift by displacing more the same amount of swept volume in a shorter distance because the wing is more tilted in to the wind. The lift force is still the same, therefore the drag is increased because the airfoil plane angle is steeper. The greater the drag the slower the airfoil has to go to lift the aircraft because it is displacing the required volume of air in a shorter distance because of t he airfoil's profile to the direction of travel. There is a maximum profile where the airfoil is a steep as it can be before drag force falls off. This point is where the geometry of the airfoil aligns congruent to the wind direction. This same force of effect acts on the propeller only the direction of travel of a propeller is in a circular direction. Still, the same variables act, only in a circular direction. There is no "optimum" drag coefficient per aircraft. The maximum applies to every aircraft. Reducing the maximum increases the required speed of the airfoil to sustain lift at takeoff because the drag is reduced and therefore the distance required for the required volume of air force to be displaced at speed is increased. The required limits of tolerances for the airfoils to displace the required lift force also is dependent on the volume of the airfoils. The volume of the airfoils also reduces the required distance to sustain the required swept volume, however the lift force is unchanged. If an aircraft weighs 340 pounds it still needs 340+ pounds of lift while still sustaining the shortest possible swept volume displacement distance. Doubling the airfoils displacement to shorten the swept volume displacement distance by half while only having half of the required thrust necessary to sustain lift will not create enough lift for the aircraft to fly. There must be as much plus thrust for the aircraft to fly as the aircraft weighs. No fractional part of the aircraft weight in thrust will develop enough lift force for the aircraft to fly. All aspect variables are coefficient and in proportion and are aligned with no tolerance.

340+ pounds is 340 pounds plus drag.Remember, adding drag is adding ++ and - - to drag which is both perpendicular to the direction of travel.

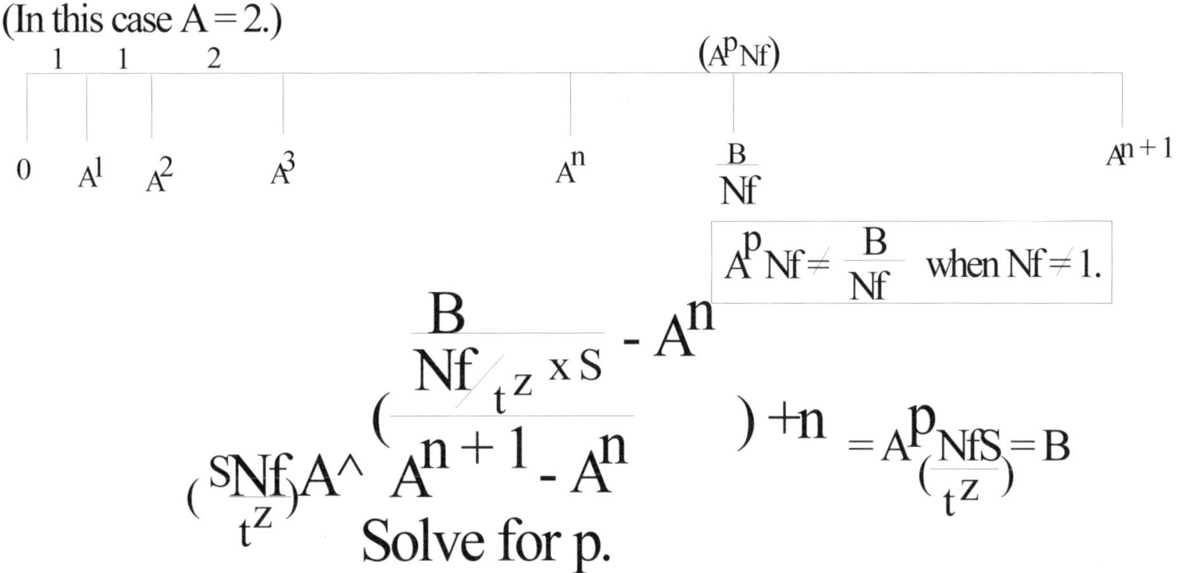

n = an integer whole number.

N = density in lbs./cu.ft.

f = volume in cu. in.

B = Force in ounces

S = oz./lb. in base 2

A = base, 2.

$t^z = 12^3$ cubic inches/cubic foot in base two.

S = some unknown quantity in any other base.

Once again the exponent power function of the calculator is causing an error.

12^3 is used in base 2. What other units of measure would be used for other bases?

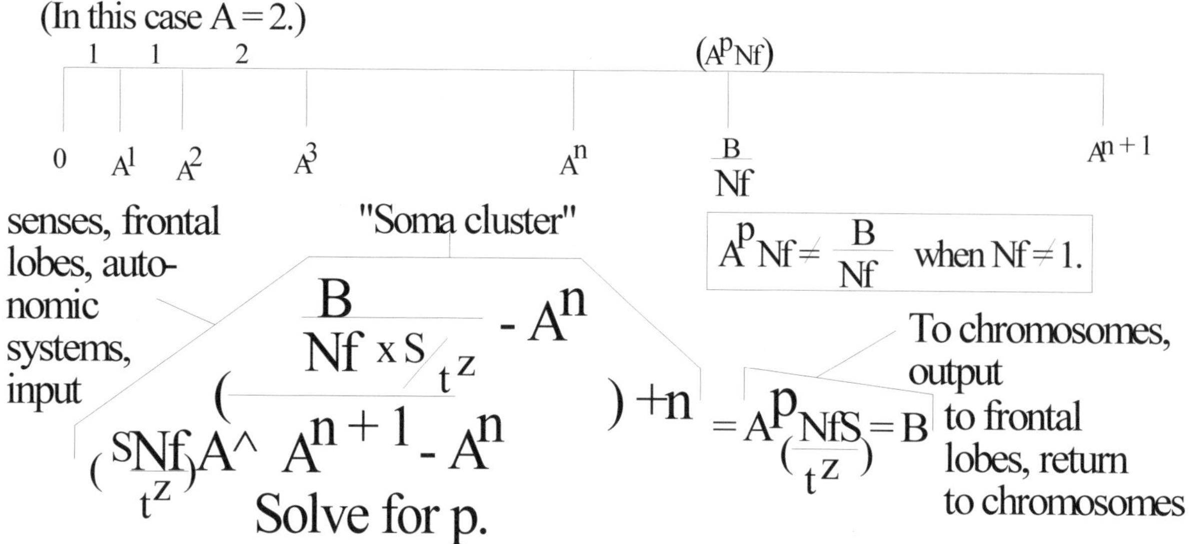

S = some unknown quantity in any other base.

Once again the exponent power function of the calculator is causing an error.

n = an integer whole number.

N = density in lbs./cu.ft.

f = volume in cu. in.

B = Force in ounces

S = oz./lb. in base 2

A = base, 2.

$t^z = 12^3$ cubic inches/cubic foot in base two.

12^3 is used in base 2. What other units of measure would be used for other bases?

Bldg

Btrlg

Ttrlg

Tldg

Em=0

Em=0

AA'

BB'

Em=0

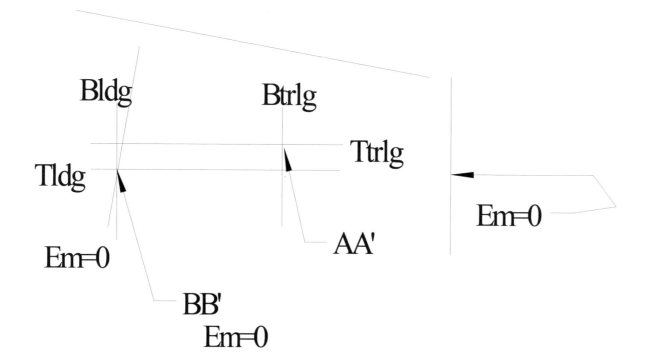

Output rear wheel and tire with central impeller.

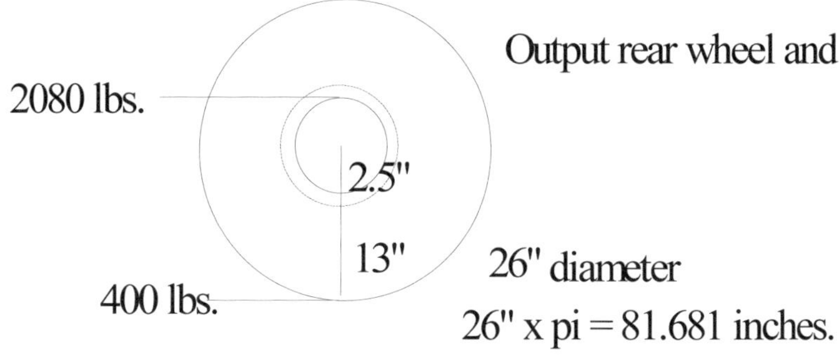

26" diameter

26" x pi = 81.681 inches.

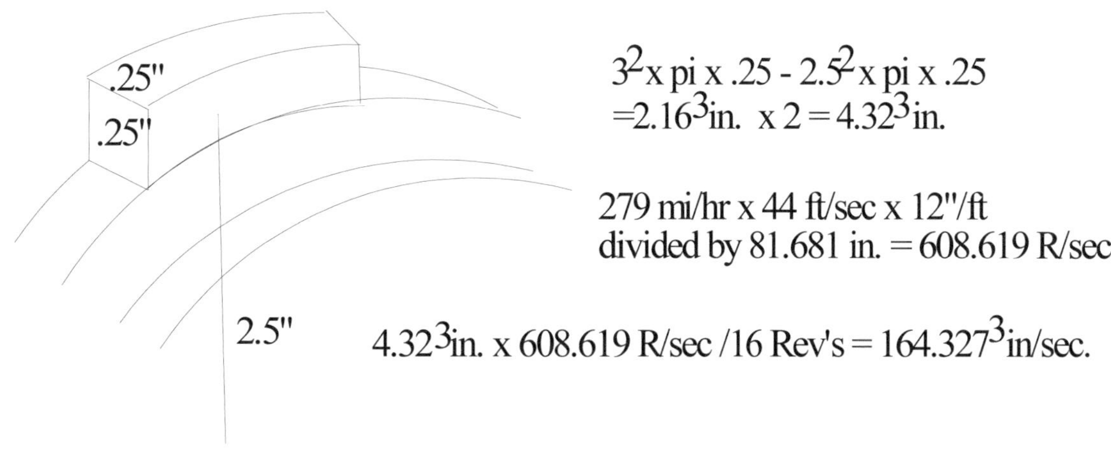

3^2 x pi x .25 - 2.5^2 x pi x .25
=2.16 ^{3}in. x 2 = 4.32 ^{3}in.

279 mi/hr x 44 ft/sec x 12"/ft
divided by 81.681 in. = 608.619 R/sec.

4.32 ^{3}in. x 608.619 R/sec /16 Rev's = 164.327 ^{3}in/sec.

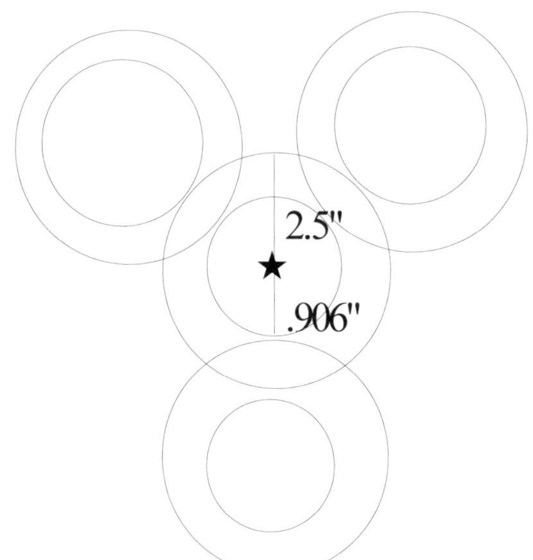

$(2.5\ \text{in}^2 \times \text{pi} - .906\ \text{in.}^2 \times \text{pi}) \times 3 = 6.99^2\ \text{in.}$

$164.327^3\ \text{in./sec} / 6.99^2\ \text{in} / 2 = 11.754\ \text{in.}$

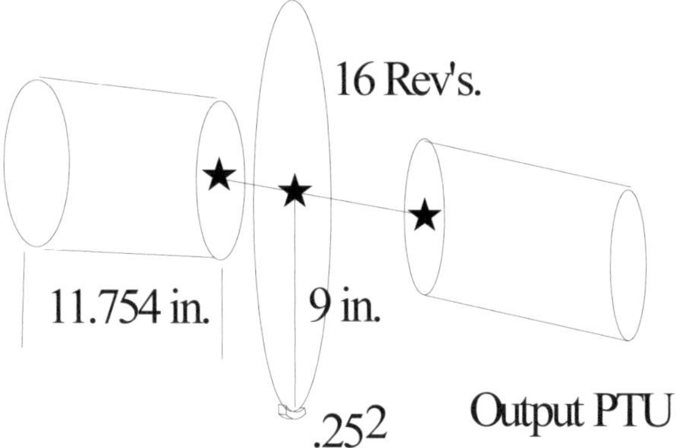

$(9.5\ \text{in}^2 \times \text{pi} \times 2.5 - 9\ \text{in}^2 \times \text{pi} \times 2.5) \times 16\ \text{Rev's} =$
$116.239^3\ \text{in.} / 6.99 / 2 = 8.315\ \text{in.}$

2080 lb. x 1.25 x 1.25 / 9 / 6 = 60.185 lb
operator's force at the crank.

Operator's force will be half since the crank ratio is 2/1, two crank revolutions per revolution of the input PTU.

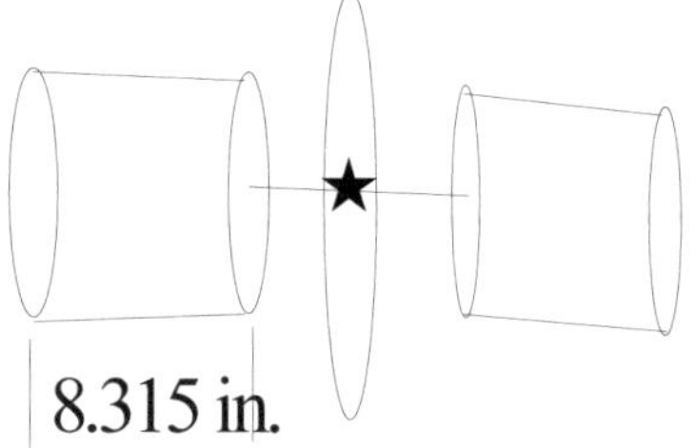

Input PTU

These are the specifications for the gears and pinion in the PTUs:
Dividing Head necessary,

2.8 Large diameter
1.937 small diameter
@30 degrees

Pinion #8 cutter
depth .432
20 degree pressure angle

4.15 Large diameter
2.25 Small diameter
@24 degrees

gear #6-1/2 cutter
spaced 3.063
depth .450

clearance depth of gear + 2 x .030

With the lift blade vanes of the human powered helicopter only slightly pitched still the applied load is 2048 lbs and the drag force is minimum. Add the lift horizontal force plus the drag force = the total force applied at the lift blade impeller blade vanes and the calculations for revolutions at the lift blade impeller will be greater revolutions than before when the pitch of the lift blade was maximum to reduce the lift blade output revolutions. Although the applied force is slightly greater to the impeller blade vanes and the rpms are greater, the result is that if the blade were heavily pitched the sum of the lift force plus the drag force would be too great to apply minimum fluid volume to the Power transmission unit meshing gear teeth volume circuit of revolution (mgtv c/r) and with the pitch of the lift blade being slight the applied force sum of lift plus drag is minimum and the fluid flow volume to the lift impeller is the least possible so the power transmission units volumes are the smallest possible. Decrease the pitch, increase the lift blade revolutions per two cranks, increase the flow volume to the output power transmission unit lift impeller, and reduce the applied force to the operator.

It is surmised that the angle of the pitch of the lift blade should be 1/u. u being the u of 2^u.

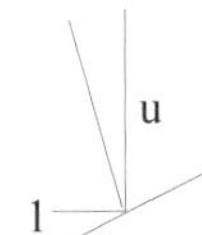

Alignment of the flywheel size with respect to its impact reaction force limit:

261 pounds
2 crank revolutions per second
gears ratio
128 crank revolutions per minute
261 / gears ratio x2 = impact reaction force

Although the actual arrangement is the opposing ratio:

1
u

1 = drag
u = 2048 lbs.

A
B o h A'
w
B' a

(lbs.) BB' = h

h sine w + h cos w = thrust (lbs.)

$$CD2^{\wedge}\left(\frac{\frac{BB'}{(CD/12^3)} - 2^n}{2^{n+1} - 2^n} + n\right) \text{sine } w +$$

$$CD2^{\wedge}\left(\frac{\frac{BB'}{(CD/12^3)} - 2^n}{2^{n+1} - 2^n} + n\right) \cos w = \text{thrust (lbs.) of vessel body volume.}$$

The necessity to have to calculate all the vectors for all the subsegments is negligible; all you will have to evaluate is the final result from the geometry of the entire vessel body BB' to AA' with respect to the angle w resulting from the equal volumes opposite perpendicular at AA' and BB' (Solid body geometry).

$$Cxbvv\ 2^{\wedge}\left(\frac{\frac{BB'}{(C \times bvv / 12^3)} - 2^n}{2^{n+1} - 2^n} + n\right) \cos w \cos v +$$

$$Cxbvv\ 2^{\left(\frac{\frac{BB'}{(C \times bvv / 12^3)} - 2^n}{2^{n+1} - 2^n} + n\right)} \text{ sine w sine v} = \text{thrust}$$

bvv = blade vanes volume of turbine blade or propeller in cubic inches
C = air density in pounds per cubic foot
BB' is in pounds
Thrust is in pounds

Sine v is the angle of association made with the first thrust of vessel body volume made with angle w applied to the blade vanes hypotenuse and its adjacent side. The original thrust of vessel body volume applies to the hypotenuse of the blade vanes applied load of what would be WR (wind resistance) bearing on the aircraft at h sine w + h cos w = thrust (lbs.).

Remember that the calculator will not calculate the base exponent solution correctly in accordance with the rules of fluid mechanics of inertia.

Rocket motor equations:

$$\frac{2^{21}\text{ pounds rocket exhaust force } \times \text{ port area or concuit area}}{\text{internal area of the rocket nozzle, } ^{2}\text{in.}}$$

= applied force to port area or conduit area (B), lbs.

$$\frac{\text{(denom.)}\; 2^{\frac{B}{\left(\frac{\text{Exhaust force, lbs, } 2^{21}}{\text{internal volume of the rocket nozzle, } ^{3}\text{ft.}}\right.}} \times \text{ displacement of turbine gallery, } ^{3}\text{in.}}{12^{3}\; ^{3}\text{in} / \,^{3}\text{ft.}}$$

$$\frac{\frac{-2^{n}}{2^{n}}}{2^{1}} + n = \frac{2^{r} \times \text{denom.}}{2^{1}}$$ = optimum applied force to the turbine gallery, lbs.

To apply throttle it must be possible to reduce the density of the exhaust force in the conduit to the turbines by allowing less exhaust in the conduit volume thereby reducing the exhaust's density by causing the gas to expand in the conduit volume. The throttle should be closed and the rocket motor idle applied output force should sustain or be equal to the weight of the rocket plus its occupants.

First Multiplier →

applied force to the turbine gallery

(exhaust force in pounds, 2^{21} / internal volume of the rocket nozzle, cubic feet. x throttle closed area / throttle full open area x turbine blade vanes volume, cubic inches

$\frac{-2^{n}}{2^{n}} + n)$ x desired turbine revolutions to sustain port area equal and opposite reaction force

x volume of turbine blade vanes x swept volume of turbine vanes / turbine vanes volume

+ clearance volume in the turbine gallery x (first multiplier)

= applied cubic volume to the turbine gallery to sustain idle.

Now this is the problem: what is the velocity of a volume of air when its simultaneous equal and opposite reaction force at impact is equal to its weight?

It would seem to hold true that different densities have different velocities at which their impact reaction force is equal to their weight.

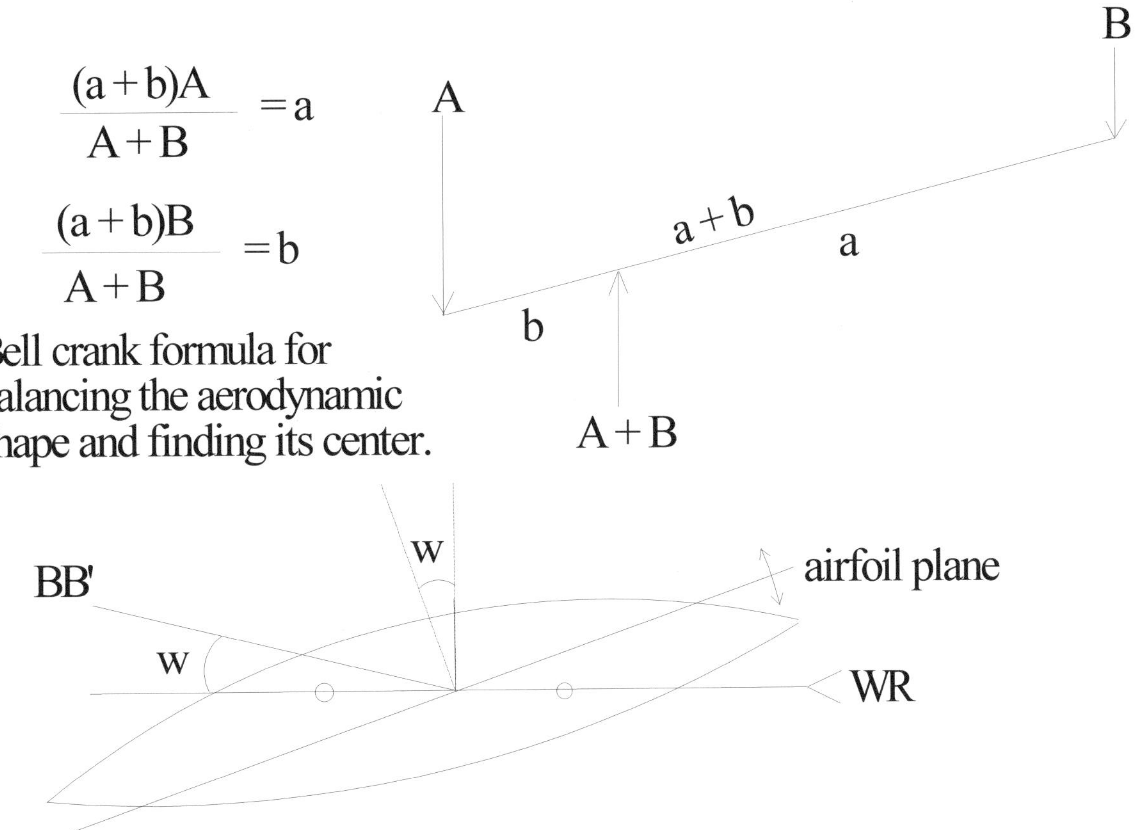

Areas geometry align their geometric centers at circles on WR. Airfoil plane pivots at lines intersection to align opposite geometric centers.

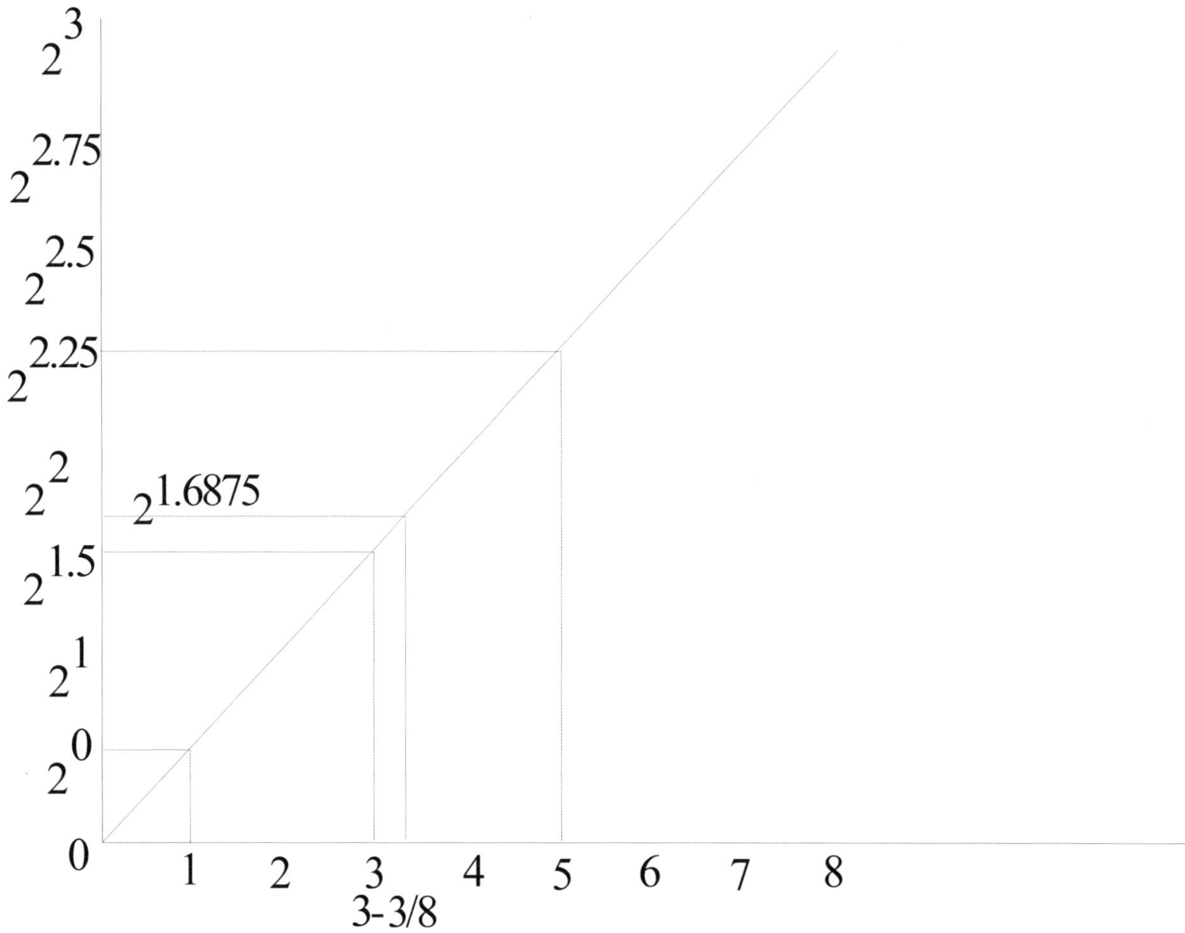

2^3
$2^{2.75}$
$2^{2.5}$
$2^{2.25}$
2^2
$2^{1.6875}$
$2^{1.5}$
2^1
2^0
0
1
2
3
3-3/8
4
5
6
7
8

Bird's Inertia
(air density x displacement of bird
swept volume at NSL
one displacement volume) x bird's inertia = impact reaction force of bird @ NSL

displacement distance of bird @ NSL
inertia of displacement of bird at air density @ NSL
inertia of displacement of bird x air density at rest = one unit displacement for one unit volume of the bird

Bird's inertia at rest x inertia of displacement of bird at air density @ NSL
inertia of displacement of bird x air density at rest

= impact reaction force of material bird @ NSL

Impact reaction force @ NSL
(inertia of displacement of bird at air density @ NSL
inertia of displacement of bird x air density at rest = impact reaction force of material bird at 1 unit displacement distance, should equal 1.

$$\frac{\text{Time the transit of units of displacement volume @ NSL}}{\left(\dfrac{\text{inertia of displacement of bird at air density @ NSL}}{\text{inertia of displacement of bird x air density at rest}}\right)} = \text{rate at 1 unit displacement distance on take off}$$

Increments of rate between different species of bird is not equal, on the ground or at any altitude.

The bird displaces the swept volume divided by * which is equal to 1 unit coefficient cubit of one bird inertia.

* = the denominator in the above equation.
NSL = normal speed @ lift.
The initial moment of applied equal and opposite reaction force to the bird displacing over the coefficient fraction of distance is equal to 1 unit cubit equal to one unit of birds inertia. The bird inertia (moment of equal and opposite reaction force at NSL is equal to the inertia of the bird x *.

The bird initially starts out at 0 displacement inertia, goes to 1, goes to 1 x 2^p. * may be referentially p. The displacement of the bird also starts out at 1 inertia and goes to 1 x 2^p .

Impact reaction force of bird @ 1 unit displacement should equal one volume of the bird's displacement in inertia @ air density.

One swept volume of the bird sustains NSL.

$$\frac{\text{One swept volume}}{*} = \text{swept volume at rate on take-off, start from 0 rate.}$$

2^p is corelevant with respect to the denominator:

$$\left(\frac{\frac{\text{inertia of displacement of bird at air density @ NSL}}{\text{inertia of displacement of bird x air density at rest}} - 2^n}{2^n}\right) + n = 2^p$$

This is an exercise. Should it prove true then more power may be applied.

22 August 2014: Although I am not sure, it seems that the closer BB' is to AA' "horizontally" the more weight the aircraft can carry, also the more vertical the line is between AA' and BB'. Whether or not the aircraft will fly with the line angle this steep I'm not sure.

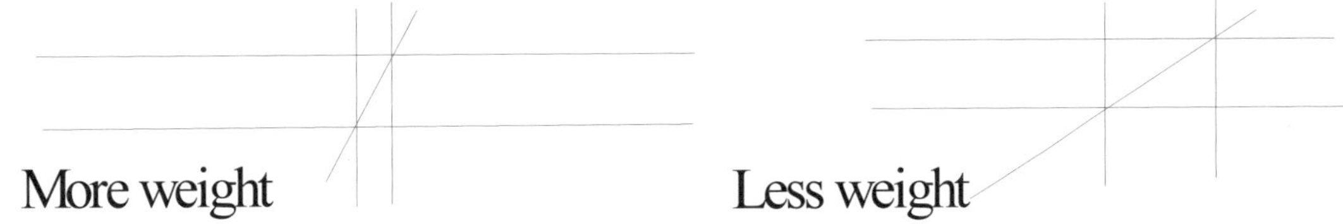

22 August 2014: Although I am not sure, it seems that the closer BB' is to AA' "horizontally" the more weight the aircraft can carry, also the more vertical the line is between AA' and BB'. Whether or not the aircraft will fly with the line angle this steep I'm not sure.

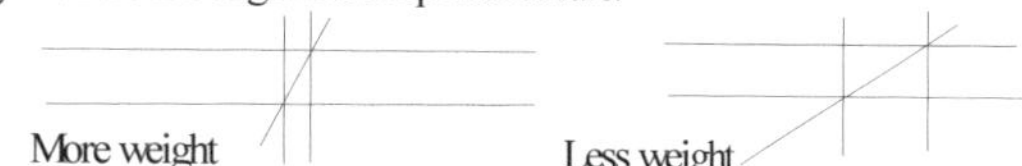

However, if the aircraft in both scenarios weigh the same the less weight scenario requires a lighter counterweight than in the more weight scenario because the lever arm is longer, the fulcrum is at BB'. Therefore, the more weight scenario aircraft must be lighter for the counterweight to be the same weight as the less weight scenario.

These scenarios have not been tried or proven in this capacity at this time. Also, the steeper angle scenario requires less speed to attain to flight than the less steep angle scenario. Also, the less steep angle scenario may fly farther at higher speed than the less steep angle scenario. The steeper angle scenario must therefore have more "lift" or shall we say drag since lift is parallel to the trajectory.

Constructing the steeper angled (AA' - BB') vectored diagram would mean the aircraft would be the lighter aircraft.

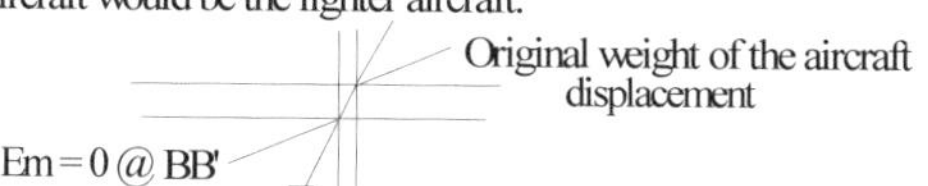

Constructing the steeper angled (AA' - BB') vectored diagram would mean the aircraft would be the lighter aircraft.

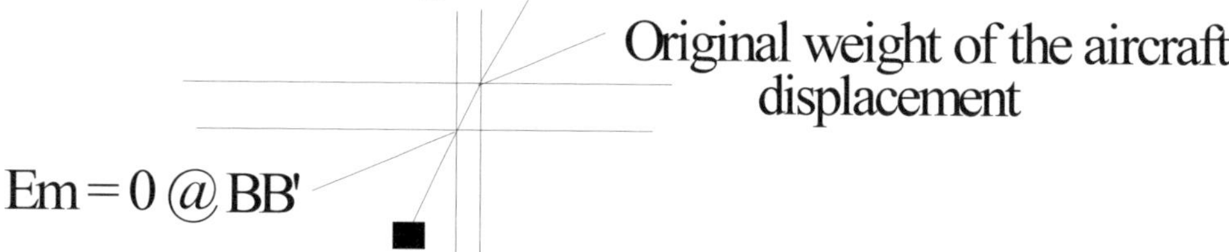

Experimentation to find the most desirable angle would require shaping the aircraft so the bisector planes align the angle at the desired angle, say: 30, 45, 60, and 75 degrees and test flying each model aircraft to see how each one flies. Shaping the aircraft to align the bisector planes to the desired angle requires aligning BB' with AA'. This may require some considerable three dimensional shape changes. The location of the wings, and the span of the wings are the most adverse of allowances to be considered to make these changes.

Making the distance between AA' and BB' as short as possible also allows for the lightest possible counterweight.

The desired angle may be different for each individual aircraft. It is possible that a conformable trigonometric ratio which is consistent with variables consistent with values from the aircraft variables will derive a ratio that will produce the desired angle. e.g. sine >= 1/p wherefore, the sine of the angle is different for individual aircraft types.

Working the equation to solve for p and using the value for n in the exponent whole number place, another possible equation which may be used for the angle is: 90 degrees minus inverse Sine of the angle (>) times the quantity (n divided by p) equals the resulting angle.

Now where n is applied n is a random anomaly and is observed as an obvious non-element in the consistent values of the aircraft's consistent make-up of material facts. n is just the number used to find p in the formula and to increase the angle from 1/p to n/p. n is just the base 2 exponent integer whole number when calculating the formula for fluid mechanics of inertia. p is the desired solution once the formula is calculated.

Model plane displacement: 12.055 cubic inches
air density: .07651 pounds per cubic foot
Conversion factors: 12 cubed cubic inches per cubic foot, and 16 ounces per pound
Balsa wood: 8 pounds per cubic foot

$$12.055 \times 8 / 12^3 \times 16 = .89296 \text{ ounces}$$

$$\frac{.89296}{12.055 \times .07651 \times 16 / 12^3} - 2^n / 2^n + n = p = 6.63377 \qquad n = 6$$

$$90 - \text{Sine}^{-1}(n/p) = 25.24894 \text{ degrees.}$$

Unless of course the opposite is true and you want the angle to be as shallow as possible: then, sine > = n/p, then in this case the inverse Sine > times n divided by p = 64.75106 degrees. It all depends on the results when the plane flies.
No other variables have been determined to fill the values for the sine ratio which are pertinent in relationship with the aircraft. Any other values which may be determined may be tried.

Values should be variables used for equations so their consistency is pertinent. Other variables which may be used are universal constants.

The solution to aligning where BB' goes with respect to AA' is: when AA' is found the displacements of the swept volumes of trailing and descent swept volumes distances and their resulting hypotenuse describe the angle, and placing the wings so BB' is on the hypotenuse and aligning the trailing volume and the leading volume to be equal should make the distance from AA' to BB' be the least distance.
This will require continuous recalculation of AA' as the wings are moved and BB' is relocated as trailing and leading volumes are recalculated. As well, upper and lower volumes are recalculated to find BB'.

Simplifying these recalculations would require a computer program.

This, however, may seem simple, but the actual determination for the location of BB' with respect to AA' when the lift and drag swept volumes are equal to one (1 u.c.) unit cubit of displacement of the aircraft volume the resulting glide slope is the resultant or the hypotenuse, and BB' and the leading volume plane describing the lower volume also must describe BB' on the hypotenuse or the resultant, BB' and AA' being on the resultant the hypotenuse. Thus, the wings may not necessarily be on the hypotenuse, and the lower and upper volumes may never be equal with a gap between them. An airplane with no gap between the lower and upper volumes being equal will not fly. The wings must be placed so there is a gap between the upper and lower volumes and the trailing and leading volumes so that the swept volumes trailing and on descent vertically equal to one vessel volume describing the hypotenuse align the four volumes so their "tic-tac-toe" description aligns BB' and AA' on the hypotenuse and as closely together as possible. The reason BB' and AA' are as close together as possible is to reduce the weight of the counterweight as much as possible.

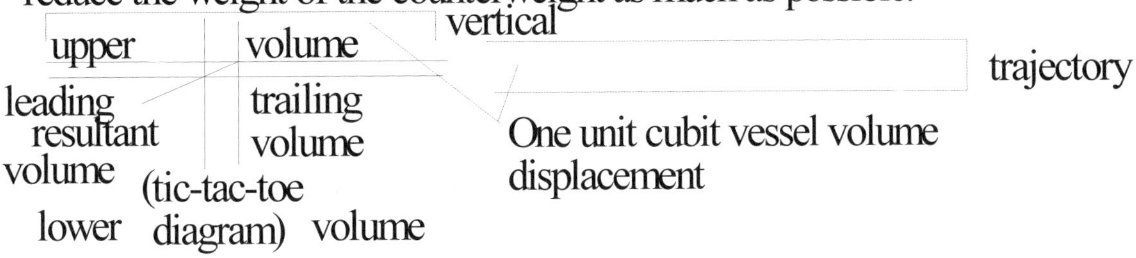

vertical

upper volume

trajectory

leading trailing

resultant volume

volume

One unit cubit vessel volume displacement

(tic-tac-toe

lower diagram) volume

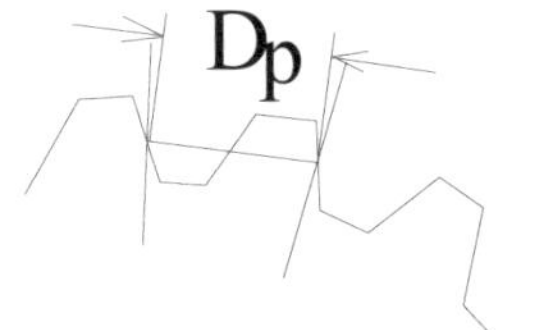

N = tooth count

$N_{(a)}$ = one gear or the other

D = basic diameter

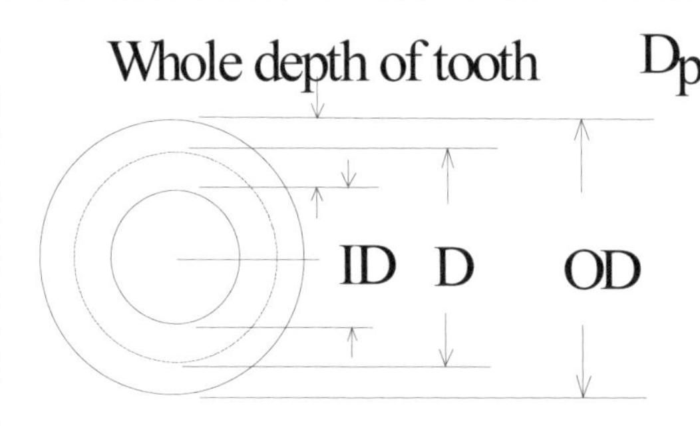

D_p = 1/8, 1/4, 3/8, etc.

This is American Standard Version.

$2 \times pi = 6.2832$

$D_p \times .6866$ = Whole depth of tooth

$$\frac{D_p N}{pi} = D$$

$$\frac{(N+2) \times D_p}{pi} = O.D.$$

$$\left(\frac{\frac{(N+2) \times D_p}{pi}}{2} - D_p \times .6866\right) \times 2 = I.D.$$

$$.6865 = \frac{703}{1024}$$

$$\frac{704}{1024} = \frac{11}{16}$$

$$\frac{(N_1 + N_2) \times D_p}{6.2832} = \text{Centers Distance}$$

$O.D._1$

$O.D._2$

.6865 may be used to avoid copyright infringment.

Add + 0.014 inches to O.D. for clearance.

...I don't know where .6866 came from.

Variables: Secondary power transmission unit = 16T + 17T
Primary power transmission unit = 6 count per side
Primary impeller @ 8 Rev./2 crank c/r & 12"r. 1/4" x 1/4" impeller blade vane profile area

How to make the variables change to obtain the optimum performance. The aerodynamic force, WR, reacting on the vehicle at F_R at optimum velocity is a value, and the evolution of the vehicle's powerplant and thus the vehicle's displacement will align at optimum limits of tolerances when the applied performance output of the powerplant by the operator (formula for Fluid Mechanics of Inertia), the geometric volume of the vehicle, and WR force all align so the Em = 0 at maximum performance: the operator is cranking at maximum applied cadence and endurance, the displacement of the vehicle's powerplant is producing the most power, the vehicle is going as fast as it will go, and the displacement of the vehicle is the trimmest it can be creating the least amount of WR force possible when WR force is maximum.

This is an equation for a mathematical genius. I do not know if I can produce an algorithm that will incorporate all these variables in to one equation. Some variables are recorded here but they may not be accurate: 2 crank c/r at L/t I applied force, Mgm/v @ applied force, minimum Mgm/v_{ptu} 140 mph, WR (minimum) = F_R maximum = unknown @ 140mph Graph:

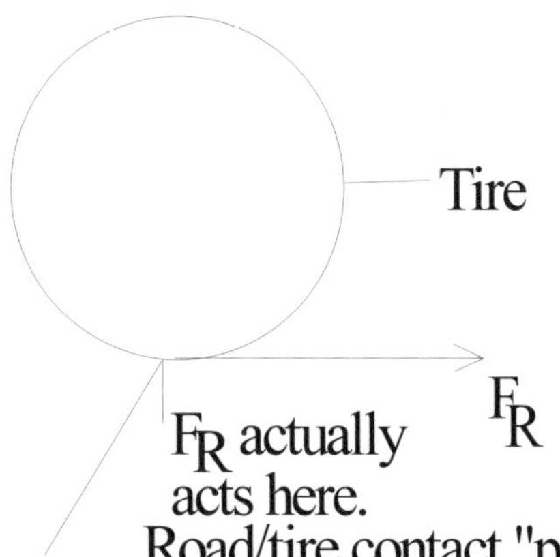

It may be beneficial to have a body on my bicycle because a body will displace fewer times in the same distance than my bicycle without a body will displace its displacement in the same distance creating a greater n value than a body will. It will be a challenge to get the least possible n value as well.

Road/tire contact "patch area".

Now for calculations: calculate the entire volume of the impeller volume plus the clearance volume, then subtract the impeller volume. Write this down. Then subtract the impeller volume, write this down. Then adding the impeller volume multiply 2 times the impeller volume. Write this down.

When calculating the fold exponent, in the denominator use the volume of the subtracted impeller volume for the fold exponent.

When multiplying the fold exponent times the volume of the impeller gallery times the revolutions use the 2x the impeller volume solution to get the answer.

Divide by (meshing gear teeth area x count of meshing gear teeth x pairs of power transmission units x rpm = depth of gear.

The reason for using the smaller volume to calculate the exponent and the larger volume to calculate the gears depth is because it results in more flow volume from the meshing gears to the impeller gallery. Actually, I'm sure that using the smaller volume to calulate the exponent yielding a bigger exponent results in a bigger product when multiplying the exponent times the larger volume times the revolutions rather than using the larger volume to calculate a smaller exponent and then multiplying resulting in a shorter gears depth.

Since there is more fluid in the gears depth therefore more volume flows throughout the clearance volume in the impeller gallery applying more aerodynamic force to the impeller blade vanes yielding more torque to the meshing gears applied to the impeller at its center gear.

It is discovered that the 2x the impeller volume yields too great a volume. Only use one impeller volume to add to the impeller and clearance calculated volume: $r_1^2 pi \times w_1 - r_2^2 pi \times w_1 -/+ (r_3^2 pi \times w_2 - r_2^2 pi \times w_2) = v_{1/2}$

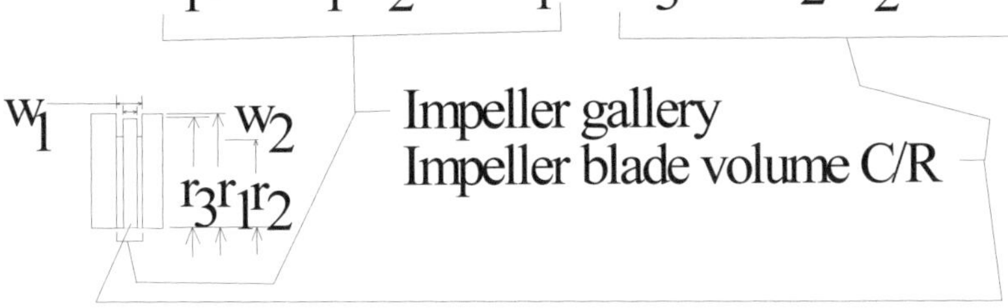

w_1 w_2
$r_3 r_1 r_2$
Impeller gallery
Impeller blade volume C/R

$$\frac{\text{exponent} \times \text{larger cubic volume of impeller gallery} \times \text{impeller rpm}}{\text{hydraulic fluid density} \times \text{count of ptu's} \times \text{revolutions per 2 crank Circuits of Revolution of ptu's center gear}}$$

= meshing gears teeth gears depth of ptu gears

$$\frac{\text{lbs. applied force}}{\frac{\text{hydraulic fluid density} \times \text{smaller cubic volume of impeller gallery}}{12^3}} - 2^n \Big/ 2^n + n = \text{exponent}$$

alternative method:

$$\frac{\text{Larger cubic volume of impeller gallery} \times \text{multiplier (e.g. 8)}}{(\text{Ptu center gear tooth area} \times \text{count of planet gears}) \times \text{count of power transmission units}}$$

= ptus gears depth.

Hydraulic fluid: anti-freeze, sewing machine oil, etc.

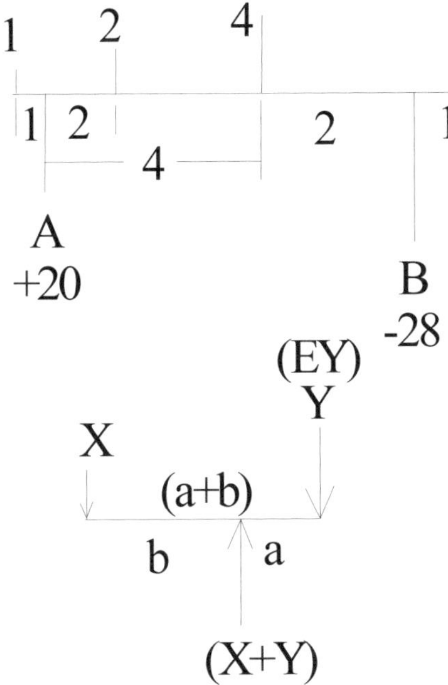

A=-1(1)+2(4)+6(2)+9(1)

B=+1(1)-2(2)-6(4)-9(1)

Find the fulcrum two loads at a time eliminating one by one, and then the two distances remaining from the fulcrum to A & B are the percentages of the sum of the loads at A & B.

Y= 2, 4, other 1, etc. = sum of Y = EY
X= first 1.

$$\frac{X(a+b)}{(X+Y)} = a$$

$$\frac{a}{(a+b)}(X+EY) = A \quad \Big| \quad \frac{b}{(a+b)}(X+EY) = B$$

Using the same formula from the previous page, calculate the fulcrum(s):

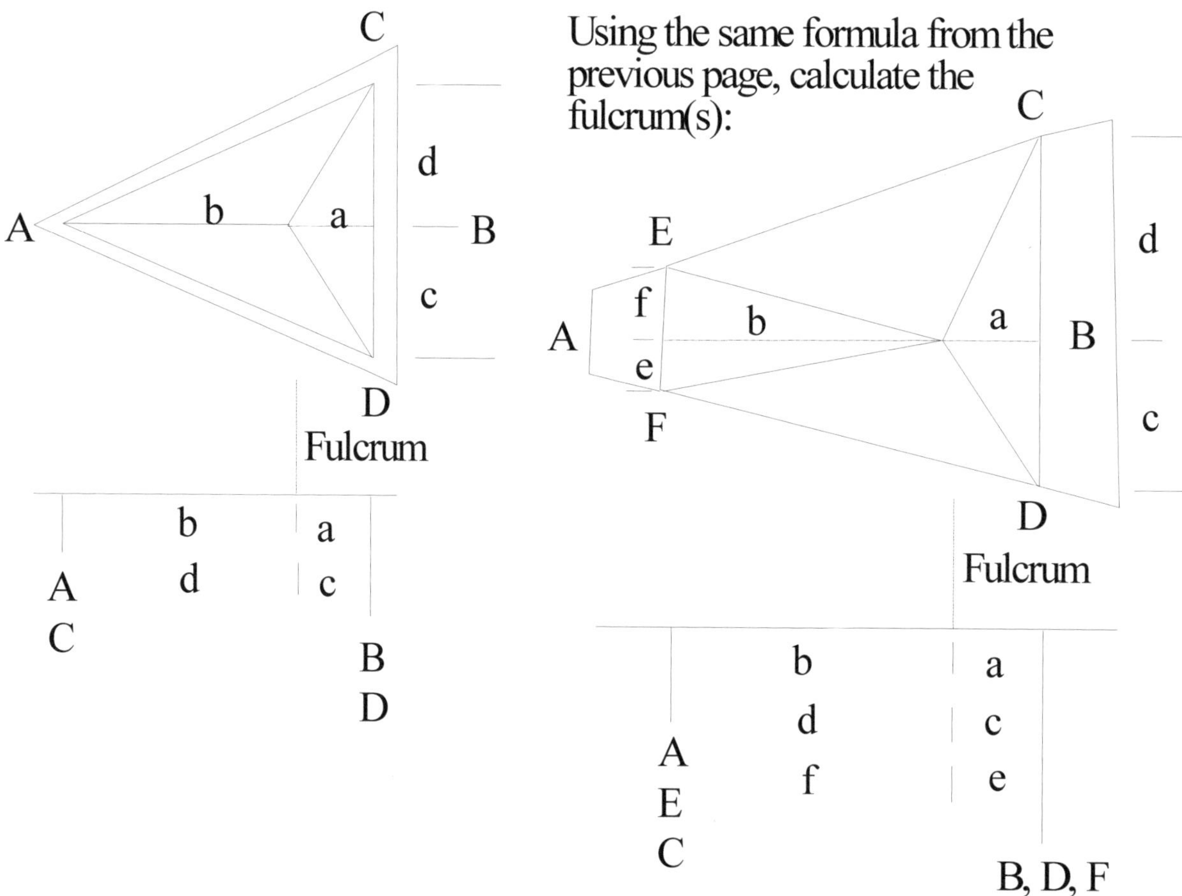

Now that I've thought about it for a while: a, shall we say, International space station has a sum density of the sum of all its densities x each quantity of each densities displacement, in consistent units, equal to a force. The ISS has one complete volume by which the sum of the former is divided by the one complete density. Space vacuum has no density. The solution constitutes the density of the ISS. The static balancing of each force for each individual density finds the center of density for the sum of densities. Calculate the geometric center of the volume of the ISS. The center of densities and the geometric center of the volume of the ISS will be different, or non-congruent. As the ISS is orbiting the geometric volume center will take its place away from the pull of "gravity", and the center of densities will take its place towards the pull of "gravity" in orbit. Apply one concentric revolution to the ISS per orbit at the center of densities on a plane perpendicular to "gravity". In this case Earth is isolated from other gravity sources. If the orbit is stable at altitude the ISS should remain at altitude and have a consistent attitude to "gravity" as it rovolves once per orbit. However, the result presently (2015a.d.) will most likely not put the ISS in a desirable attitude, however the use of attitude control jets will be minimized. Atmosphere has consistent (+ +/- -) densities and its/their respective geometric volume center(s) and is to be included in calculations with the sum of the densities.

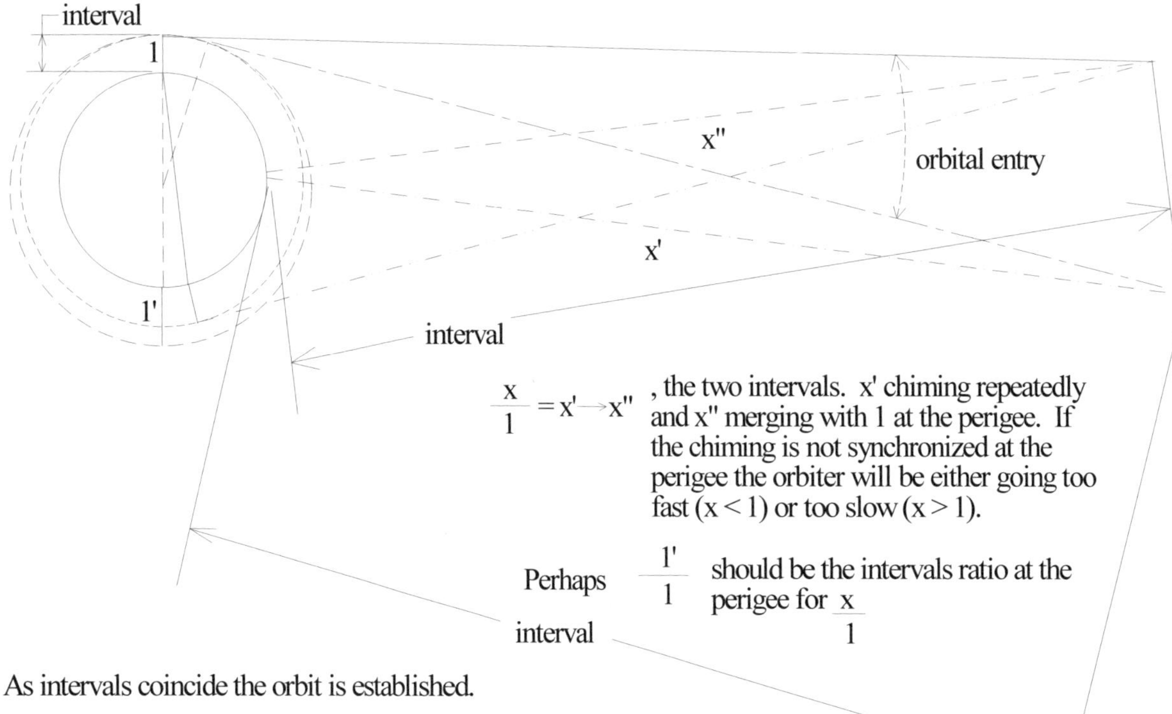

interval
1
x''
orbital entry
x'
1'
interval
$\frac{x}{1} = x' \rightarrow x''$, the two intervals. x' chiming repeatedly and x'' merging with 1 at the perigee. If the chiming is not synchronized at the perigee the orbiter will be either going too fast ($x < 1$) or too slow ($x > 1$).
Perhaps $\frac{1'}{1}$ should be the intervals ratio at the perigee for $\frac{x}{1}$
interval

As intervals coincide the orbit is established.

A computer would probably have to synchronize the chiming between x and 1 as the approach becomes increasingly closer to 1 at altitude.

and as x becomes perpendicular to trajectory.

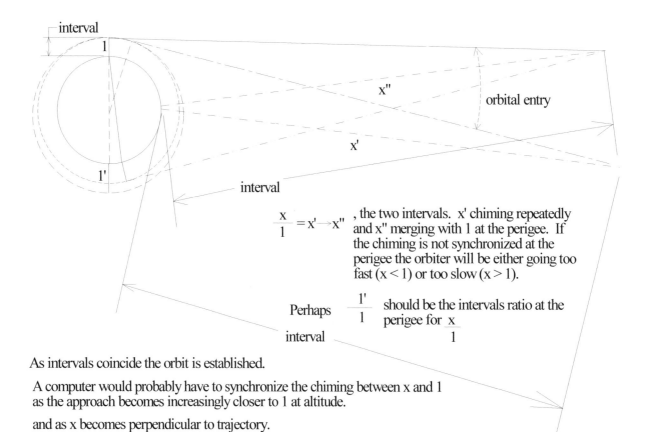

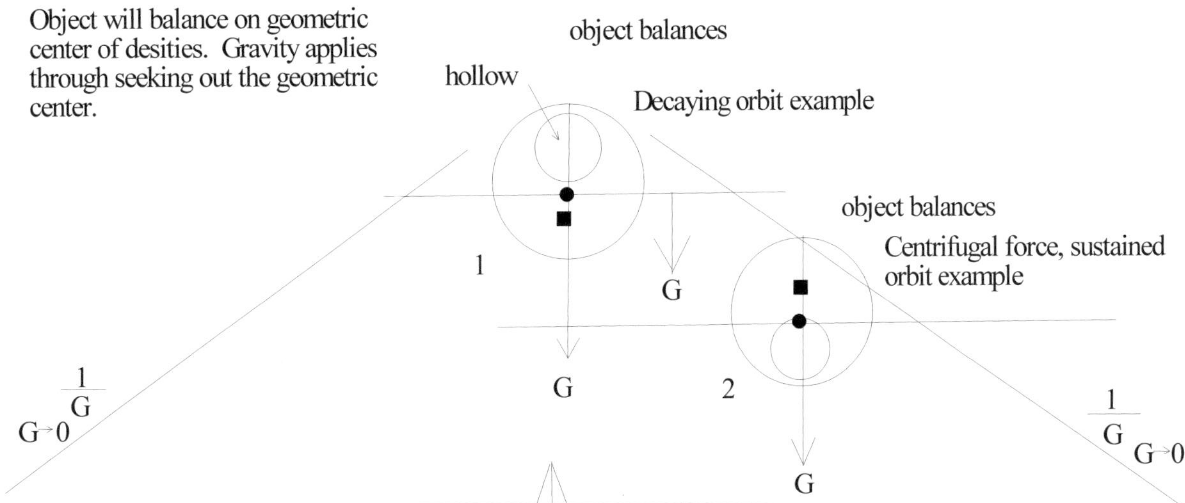

Experiment: evacuate the hollow and rebalance. Does the evacuated hollow constitute volume? Does gravity find a new geometric center if the evacuated hollow does not constitute volume?

Example 1 has an orbital velocity slower than example 2. The variation in velocity my not be very significant.
The orbiter will roll on orbit if its velocity is too low, increasing velocity will stabilize the roll to a "top heavy" attitude in orbit: Stable orbit.

Planet

The variation in velocity between two bodies of equal volume is coefficiently proportional with respect to their two attitudes on the orbit and the ratio of their densities times the one's velocity.

If they are equal density their velocities ratio is incomparable. No solution is yet found.

G

Funny, that the Earth doesn't have any planetesimals and neither does the moon and so many Eons. It would seem that that prospect of probablity is too extreme for their to be any.

The moon would therefore be orbiting having its center of density on the outside from the Earth because it has always got the same face towards the Earth, and therefore is in a stable orbit.

How astronomically improbable is that? The moon could very well be a dead star.

Being a small orbiter in a stable orbit there will likely be a rocking motion in orbit as the orbiter processes the cyclonic orbit. This rocking motion is normal and will initiate at the minimum velocity for a stable orbit although the rocking motion may be more pronounced at minimum velocity for a stable orbit. Acceleration somewhat may reduce the exaggerated rocking motion of a minimum velocity stable orbit but will not quell it altogether.

The more excentric the density center is from the geometric center the less likely there will be a tendency for an exaggerated rocking motion.

It may be possible to align an orbit with an orbiter so that the perigee pendulums back and forth on the sun side of the planet the orbiter is orbiting. The farther out from the planet the orbiter is the more sensitive the alignment will be.

Shear area formula:

$$(pi \times r_2^2 - pi \times r_1^2) \times \text{shear force of desired material in lbs/}^2\text{in.} = \text{required force in pounds.}$$

$$r_2 = \sqrt{\frac{\text{required force in pounds}}{(\text{shear force of desired material in lbs/ in. x pi})}} + r_1$$

$$r_2 \times 2 = \text{diameter.}$$

Mathematics used in the formulation of equations (Science of Mechanics):
Basic Functions
Algebra
Geometry
Trigonometry
Science of Mechanics (Fluid Mechanics of Inertis)

The Science of Mechanics appears similar to calculus but is only algebra. The Science of Mechanics implements resolving the exponent of any base with long-hand mathematics, and the resolution may be observed unlike observing the resolution for solving an exponent using a scientific calculator.

"The Formula for Fluid Mechanics of Inertia" is also a basis for analytical brain function, the likes for which how chemicals can make value changes to variables in soma clusters and develop a tangible solution to the brain creating information that an individual understands is still beyond the scope of understanding. A soma contains a variable, a soma cluster is an equation e.g. smell, or taste etc. Information enters the brain through its senses sending chemical and electrical signals to the soma clusters and the brain is continually making variable adjustments to the values in the somas and the solutions are continuous and uninterrupted. The frontal cortex also conatins soma clusters which operate in the same manner using processing of information gathered throughout ones life. The chemistry of the frontal cortex however, is beyond the scope of this text.

Human Powered Dragster

From the start this dragster is designed to reach a speed of 230 miles per hour at a cadence of 128 revolutions per minute. Approximately 40 pounds of pedal force will be applied at the cadence of 128 revolutions per minute. The dragster is equipped with a twist-grip throttle, hand brake levers at both hands, brakes, and the usual power transmission assembly that I have designed. The dragster has steering for which the front wheels are designed to steer in parallel due to the fact that the track it is designed to be tested on is only and will only be a straight strip of pavement one quarter mile long. Should the vehicle side-slip at any time, the capability of being able to direct the front wheels in the desired direction will point them both simultaneously as required, none of this cornering business is necessary on a straight track. Both front wheels are necessary for traction when steering in this case so front wheels parallel steering is required in order to assess the vehicle in the desired direction.

An applied force to the road at the contact point where the rear tires are on the pavement at the maximum applied performance is 1500 pounds when approximately 40 pounds of force is applied to the pedals at full throttle during cranking at the cadence of 128 revolutions per minute.

The differential of the vehicle is a dual set of opposing turbines. On one side of the final output impeller of the power transmission assembly “differential” are two circularly patterned turbines facing one another separated by a metal plate between them; beyond the perimeter and at the inside circumference of the blade vanes of the turbines are openings in the metal plate to allow the hydraulic fluid to circulate. One turbine draws in the circulating hydraulic fluid through the metal plate openings at the perimeter of the metal plate and forces it in to the receiving chamber of the other turbine through the opening at the center of the metal plate. The driven turbine blade vanes have a 0.0035 inch clearance from the metal plate. The metal plate is attached to the center assembly of the power transmission “differential” impeller and drive turbine. The hydraulic fluid in the receiving chamber of the driven turbine is forced out through the driven turbine blade vanes (The driven turbine blade vanes are a circular pattern of the drive turbine blade vanes, rotated 180 degrees having a center at .06425 inches from the face of the drive turbine blade vanes) which forces the driven turbine to respond. The cycle is continuous as long as power is being applied to the differential hydraulic fluid flow input by the operator. Otherwise the vehicle can coast: however, the vehicle will not reverse.

The front steering is simply a rack and pinion set-up. The operator has handlebars in the operator’s compartment to steer the vehicle. Bearings are required all around and all over the vehicle, including the power-plant assembly. Also, numerous grease seals are required, and grease nipples, and bleed valves in places to expel air from the hydraulic lines. It also needs a bunch of ¼ and 3/8 inch threaded nuts and bolts with drilled and threaded holes to match, and washers, screws, and even some holes for some shafts with E-clips to hold the shafts in place. The body panels will most probably be welded on. All the mathematical prospects have been accomplished except for the actual performance, and once again all I can do is draw and write and do arithmetic. I have no funds to pursue any furtherance of the prospects of my ideas and discoveries. Being able to afford the limits of my capability is beyond my means. In this day and age the saying should be like: "Lots of things to do and no crowd-funding website contributors will fund any of my projects."

The object of this publication is to deliberately void patenting applicability and to force the inability to file a patent application by anyone attempting to pirate, counterfeit, interfere with, or any other concept ideal of patent law which makes it possible for some other entity to file a patent application on my idea and discovery. I deliberately intend to usurp the first and original inventor status by publishing my work before any application for patent can be filed by anyone else. This is a deterrent strategy to prevent anyone else from signing a patent application for my ideas and discoveries knowing full well that they are not the first and original inventor, if that is the case.

The following image is a manually propelled crank driven human powered dragster engineered to sustain from 0 - 230 miles per hour within a quarter mile distance is less than 3 seconds. The drawing made by CAD software is full scale in the CAD drawing file: 1" = 1"

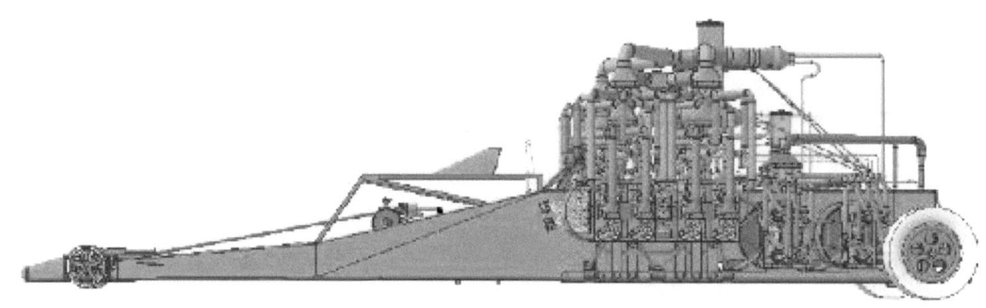

Left side view: manually propelled crank dragster

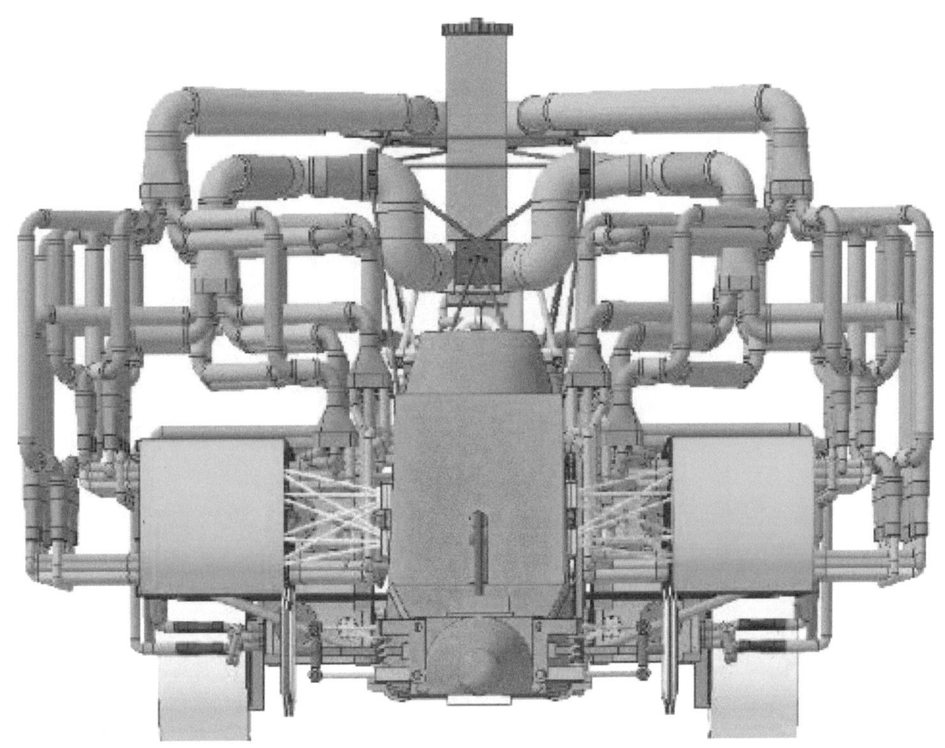

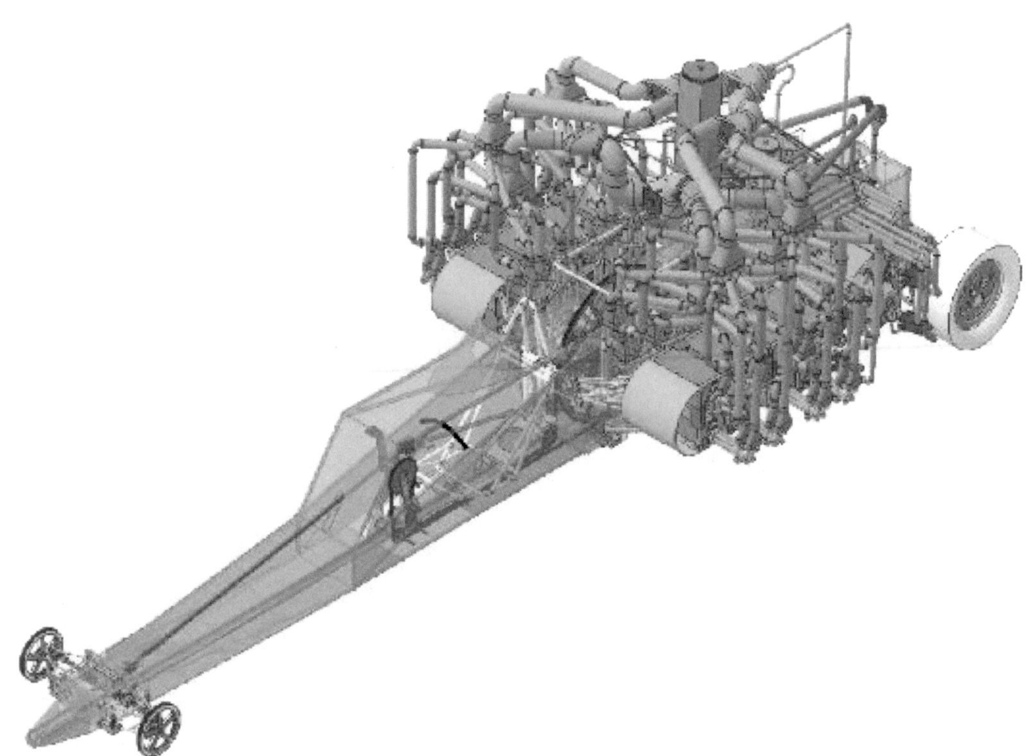

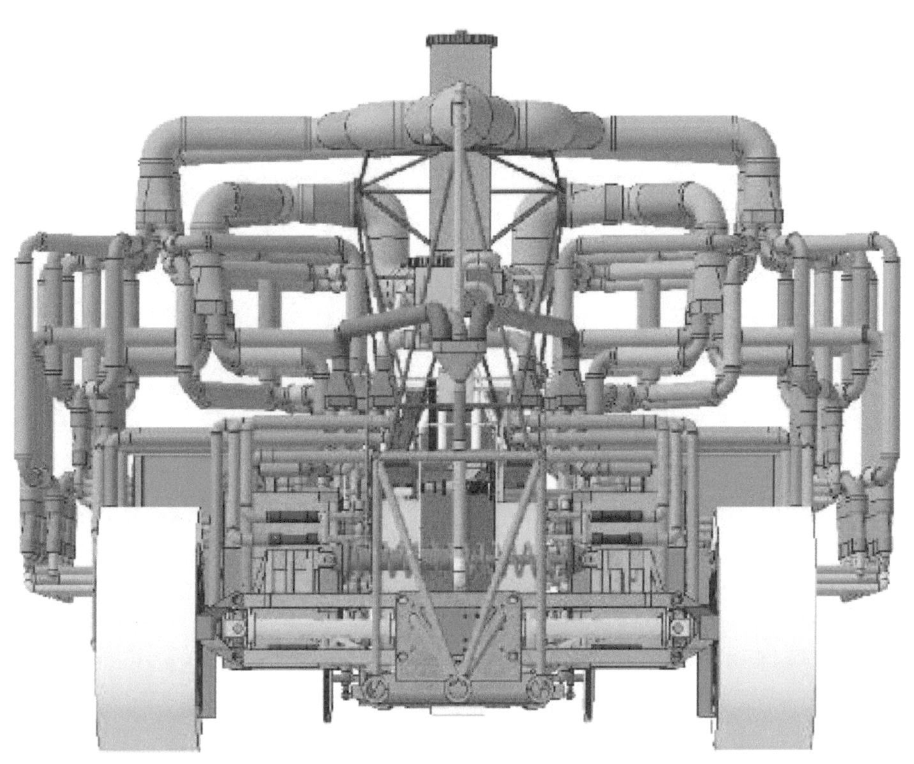

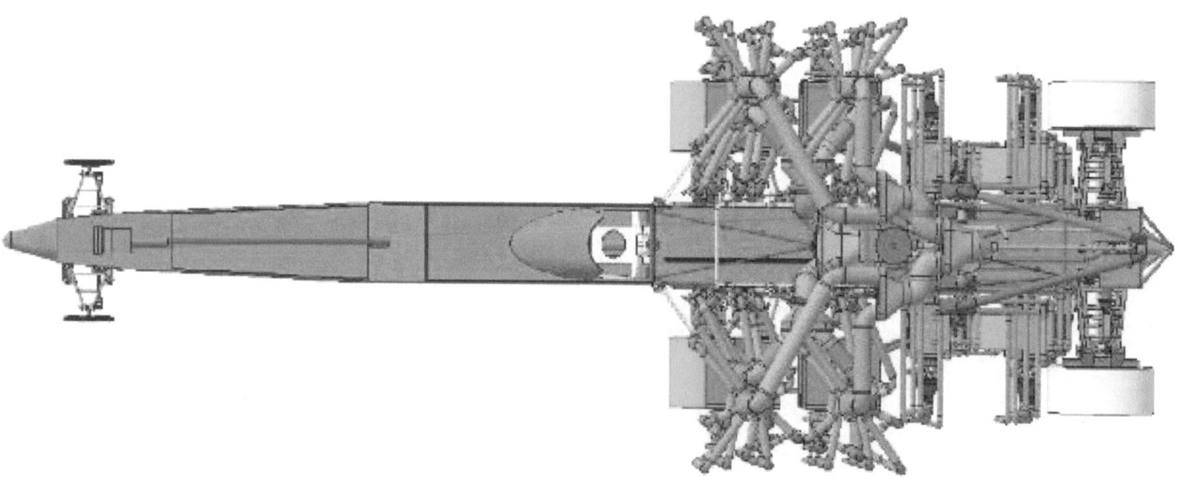

Furthering the application for the human powered helicopter: The alignment to apply minimum tolerances to parts dimensions so that any amount exceeding the maximum allowable limits for applied force to the crank would jeapordize the elastic limits put upon the applied parts for the hydraulic power-transmission system thereby reducing the weight of the aircraft to minimum. This will probably come about by rigid calculations to determine these limits in application to performance and their respective materials out of which the parts are made. This determination is left for a future cause at this time. Just getting the human powered helicopter to fly right now is first priority.

Jettisoning the weight is a lesser priority right now.

How many degrees does the flywheel rotate at 1 rpm and its impact reaction force is equal to its weight? What is the cadence? 128 or sprockets ratio inverse: max/min gears teeth. Multiply by the sprockets ratio. What is the density of the flywheel and what is its volume? What is its center radius at the median circumference of applied reaction force? Make its applied force at 128 cadence equal to the operator force required at the crank pedals:

assuming:
If 19.5 degrees rotation of the flywheel at 1 rpm develops the e/o r.f. (equal and opposite reaction force) rotation of the flywheel equal to the weight of the flywheel, then 1/120 of a minute will result in the e/o r.f. = 120 x the weight at the flywheel. The applied force at the pedal in static elastic equilibrium throughout the torques from the pedal center at the crank operator force x the crank radius / chainring sprocket raidus x flywheel sprocket radius / flywheel center of applied force = 6 degrees of rotation at the crank, develops 19.5 degrees at the flywheel. Calculate the flywheel radius and density.
Area x pi x Diameter = volume.

or use a known density and calculate the flywheel radius.

$$\frac{\text{calculated force at operator pedals x pedal axis to bottom bracket axis "radius"}}{\text{chainring sprocket radius}} \text{ x}$$

$$\frac{\text{flywheel sprocket radius}}{\text{flywheel radius at center of applied force}} = \text{product force at attempted radius.}$$

(make denominator radius attempts: 5, 12)

$$\frac{\text{density of flywheel material (lbs/ ft.}^3\text{) x calculated cubic volume (est.: .25 in.}^3\text{) at the attamped radius}}{\text{product force [at the attempted flywheel radius]}}$$

x attempted radius = required radius with known material density.

The volume of the calculated volume remains the same volume at the required radius.

$$\frac{\text{Attempted radius results}}{\text{Desired radius}} \text{ x attempted cubic volume = required cubic volume at the desired radius.}$$

Remember to use 12 cubed cubic inches / cubic foot when converting from cubic inches to cubic feet or vice versa, and either multiply or divide with it.

Remember that there are two crank revolutions in one second at cadence for optimum output, actually 128 cadence, but there aren't 64 seconds in a minute, so 128 cadence is not used here, and there aren't 256 degrees or 512 degrees in a circle etc, so using base two would probably simplify the problem but it isn't taking effect here.

pi D A = V, square root of the area (A) = side.
Known density = 165 lbs./ cubic foot
Desired radius = 7 inches
6 degrees @ crank = 6 x 52 / 16 = 19.5 degrees at flywheel   360 / 19.5 = 18. 46154
77 lbs required at crank operator's force (variable)
2 crank circuits of revolution / second
Flywheel impact reaction force product count per circuit of revolution = 360 / 19.5 = 18. 46154.

flywheel weight x 19.5 x 2 = 77 lbs. @ crank arm radius
flywheel weight = 77 / (19.5 x 2) = 1.97436 lbs. @ 7 inches radius. (static force)
1.97436 x 19.5 x 2 = 77 lbs ~~(static force).~~
14 x pi x 1 inch squared = 43.9823 cubic inches.
(165 lbs. / cubic foot / 12 cubed cubic inches per cubic foot) x n cubic inches = 1.97436 lbs. (static force). n = 20.67693 cubic inches.
Static force x 52 / 16 x 2 x 19.5 - 77 lbs. = 173.25013 lbs.
square root of ((20.67693 / 43.9823) x 1 squared inches) = .68565 inches on a side.
1.97436 x 52 / 16 x 2 x 19.5 - 77 = 173.25013 lbs. of force at 7 inches radius, which is in these calculations to be the radii of both the flywheel and the crank chainring.

Dynamic force ≛ static force x 52 /16 x 2 x 18.46154 - 77 = 159.92322 lbs.

.68565 squared x 14 x pi = 20.67693 cubic inches.

Furthering the application for the human powered helicopter: The alignment to apply minimum tolerances to parts dimensions so that any amount exceeding the maximum allowable limits for applied force to the crank would jeapordize the elastic limits put upon the applied parts for the hydraulic power-transmission system thereby reducing the weight of the aircraft to minimum. This will probably come about by rigid calculations to determine these limits in application to performance and their respective materials out of which the parts are made. This determination is left for a future cause at this time. Just getting the human powered helicopter to fly right now is first priority.

Jettisoning the weight is a lesser priority right now.

How many degrees does the flywheel rotate at 1 rpm and its impact reaction force is equal to its weight? What is the cadence? 128 or sprockets ratio inverse: max/min gears teeth. Multiply by the sprockets ratio. What is the density of the flywheel and what is its volume? What is its center radius at the median circumference of applied reaction force? Make its applied force at 128 cadence equal to the operator force required at the crank pedals:

assuming:
If 19.5 degrees rotation of the flywheel at 1 rpm develops the e/o r.f. (equal and opposite reaction force) rotation of the flywheel equal to the weight of the flywheel, then 1/120 of a minute will result in the e/o r.f. = 120 x the weight at the flywheel. The applied force at the pedal in static elastic equilibrium throughout the torques from the pedal center at the crank operator force x the crank radius / chainring sprocket raidus x flywheel sprocket radius / flywheel center of applied force = 6 degrees of rotation at the crank, develops 19.5 degrees at the flywheel. Calculate the flywheel radius and density.
Area x pi x Diameter = volume.

or use a known density and calculate the flywheel radius.

calculated force at operator pedals x pedal axis to bottom bracket axis "radius"
――――――――――――――――――――― x
chainring sprocket radius

flywheel sprocket radius
―――――――――――― = product force at attempted radius.
flywheel radius at center of applied force

(make denominator radius attempts: 5, 12)

density of flywheel material (lbs/ $ft.^3$) x calculated cubic volume (est.: .25 $in.^3$) at the attamped radius
――――――――――――――――――――――――――――――
product force [at the attempted flywheel radius]

x attempted radius = required radius with known material density.

The volume of the calculated volume remains the same volume at the required radius.

Attempted radius results
―――――――――――― x attempted cubic volume = required cubic volume at the desired radius.
Desired radius

Remember to use 12 cubed cubic inches / cubic foot when converting from cubic inches to cubic feet or vice versa, and either multiply or divide with it.

Remember that there are two crank revolutions in one second at cadence for optimum output, actually 128 cadence, but there aren't 64 seconds in a minute, so 128 cadence is not used here, and there aren't 256 degrees or 512 degrees in a circle etc, so using base two would probably simplify the problem but it isn't taking effect here.

pi D A = V, square root of the area (A) = side.
Known density = 165 lbs./ cubic foot
Desired radius = 7 inches
6 degrees @ crank = 6 x 52 / 16 = 19.5 degrees at flywheel 360 / 19.5 = 18. 46154
77 lbs required at crank operator's force (variable)
2 crank circuits of revolution / second
Flywheel impact reaction force product count per circuit of revolution = 360 / 19.5 = 18. 46154.

flywheel weight x 19.5 x 2 = 77 lbs. @ crank arm radius
flywheel weight = 77 / (19.5 x 2) = 1.97436 lbs. @ 7 inches radius. (static force)
1.97436 x 19.5 x 2 = 77 lbs ~~(static force).~~
14 x pi x 1 inch squared = 43.9823 cubic inches.
(165 lbs. / cubic foot / 12 cubed cubic inches per cubic foot) x n cubic inches = 1.97436 lbs. (static force). n = 20.67693 cubic inches.
Static force x 52 / 16 x 2 x 19.5 - 77 lbs. = 173.25013 lbs.
square root of ((20.67693 / 43.9823) x 1 squared inches) = .68565 inches on a side.
1.97436 x 52 / 16 x 2 x 19.5 - 77 = 173.25013 lbs. of force at 7 inches radius, which is in these calculations to be the radii of both the flywheel and the crank chainring.

Dynamic force = static force x 52 /16 x 2 x 18.46154 - 77 = 159.92322 lbs.

.68565 squared x 14 x pi = 20.67693 cubic inches.

If the distance along the circumference between two gear teeth is equal to an American Standard units constant increment of length, then the formula pi times the diameter divided by the number or teeth equals that increment of length, whether it be 1, or 1/2 or .5, or 5/16 etc. With algebra the formula can be manipulated with the diameter as the unknown and applying the number of teeth will vary the diameter. The American Standard increment is known as the circular pitch. Other formulas for spur gear teeth can be found in the Machinists Handbook.

As it turns out, the outside diameter of the pinion gear is 2.2281 inches and the outside diameter of the driven gear is 2.54648 inches because the driven gear has 14 teeth. The driven gear diameter is the same as the pinion gear outside diameter, 2.2281 inches, so the centers distance is one half the diameters of either gear, or 1/2 x 2.54648 + 1/2 x 1.90986 = 2.06898 inches.

You want this distance to equal .5 (see example)

.5
30.00°
Circular pitch

D

Varying the teeth count will vary the degrees and the distance with respect to American Standard Units.

$$\frac{\text{pi} \times D}{T} = \text{whole number}$$

example: $\frac{.5 \times 12}{\text{pi}} = 1.90986"$ diameter D

You have the whole number of 12 gear teeth, T.

Varying the gear teeth count will not change the whole number. D is the unknown.

Gears have three diameters. There is a diameter between where the gear teeth bottom out, and there is a diameter at the top of the gear teeth, and there is a diameter in between those two diameters. The mid diameter is where most of the initial calculating begins. It is also the diameter to determine the centers distance from. Finding the middle circumference's diameter is where I have had the most trouble starting my gears drawings.

The Machinists Handbook has the gear cutter numbers for the gear teeth count that you will need to use.

The total gear depth is how deep you should machine the gear teeth.

Human Powered Helicopter

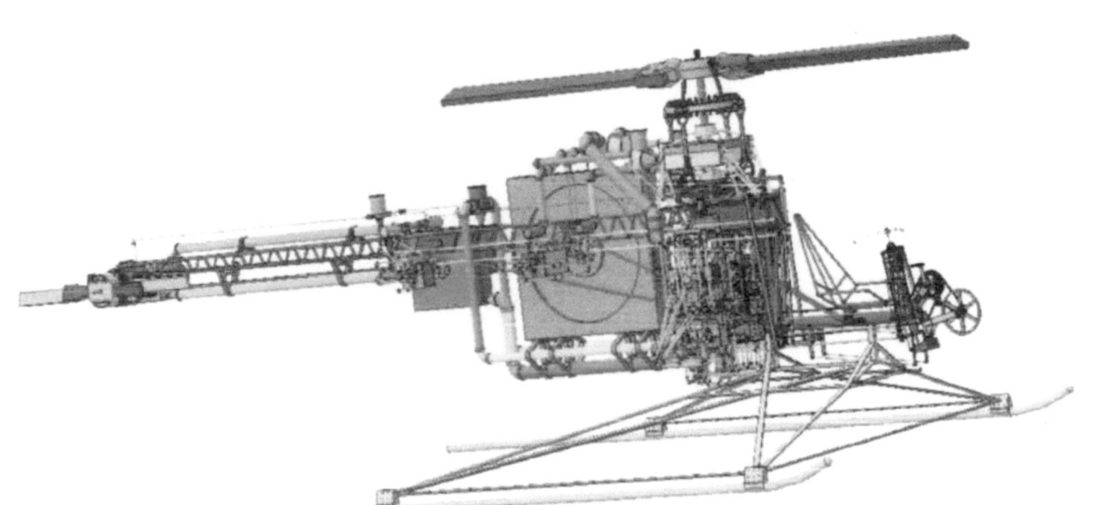

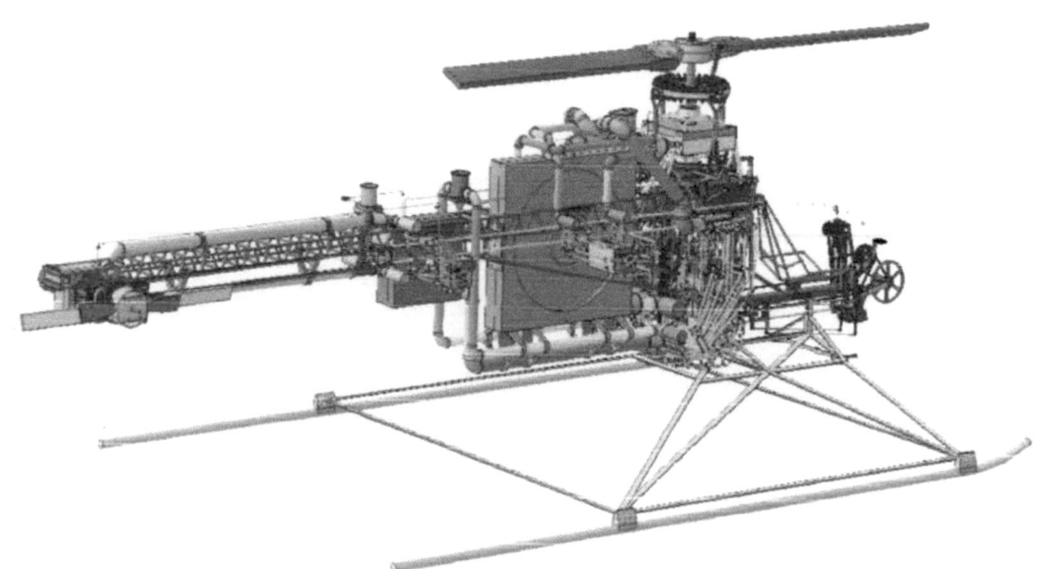

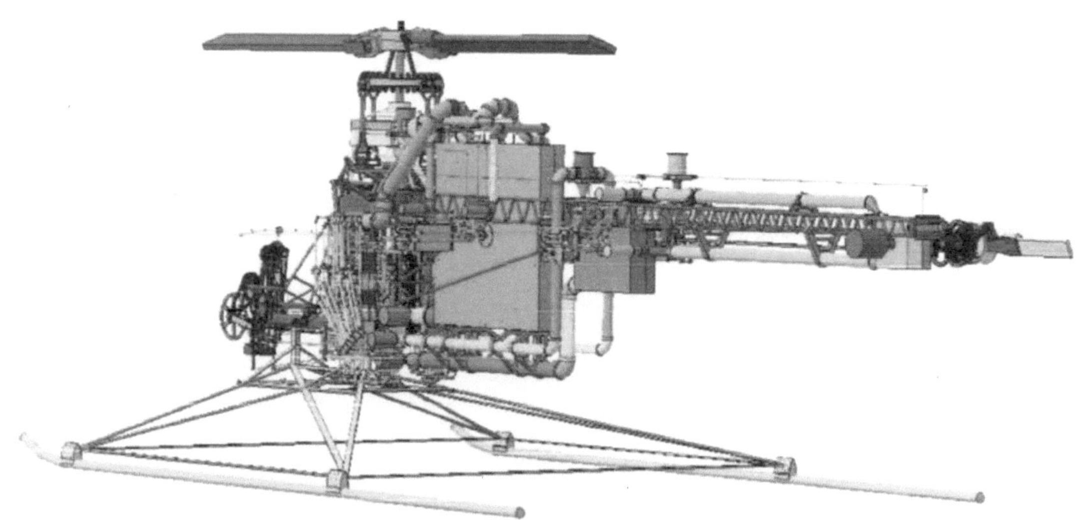

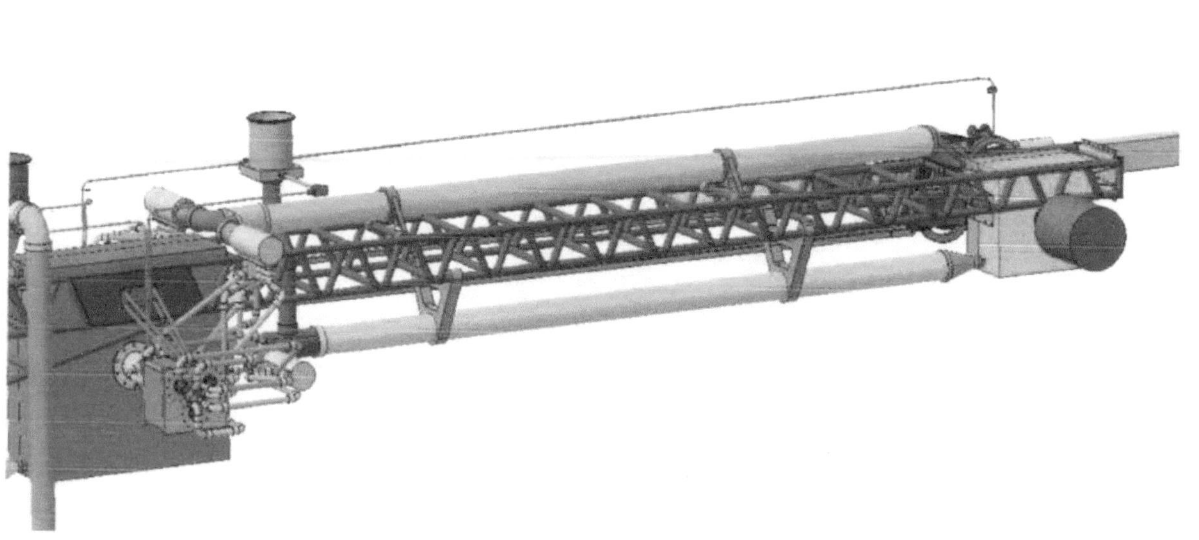

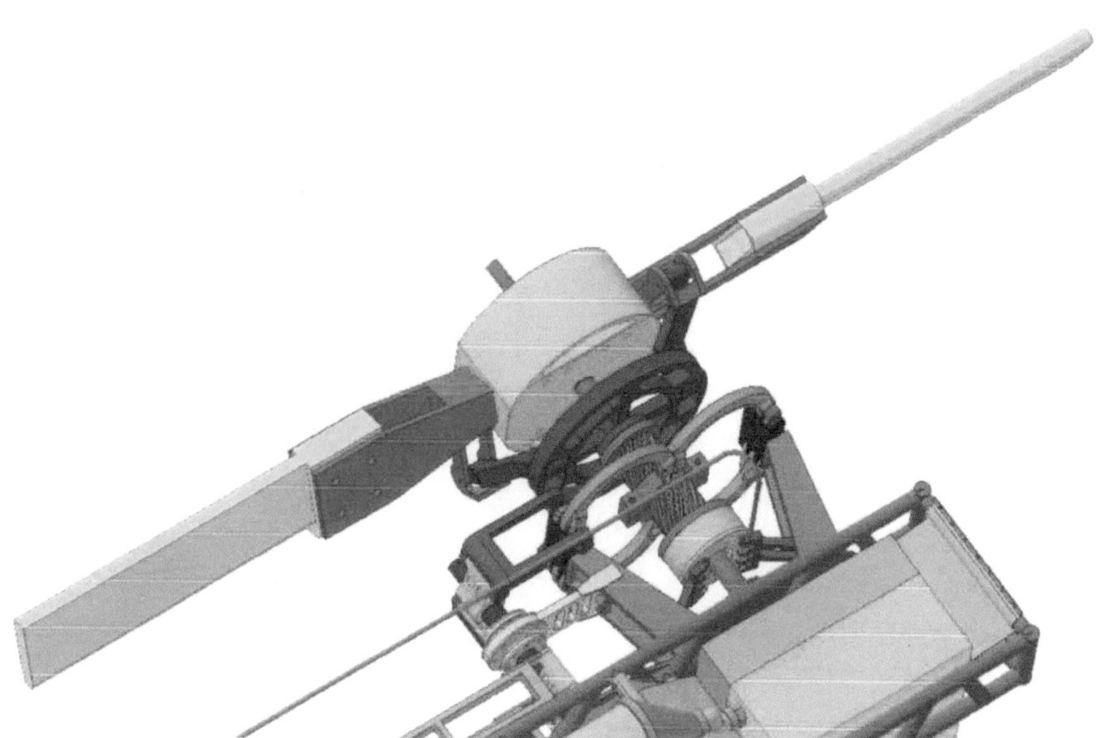

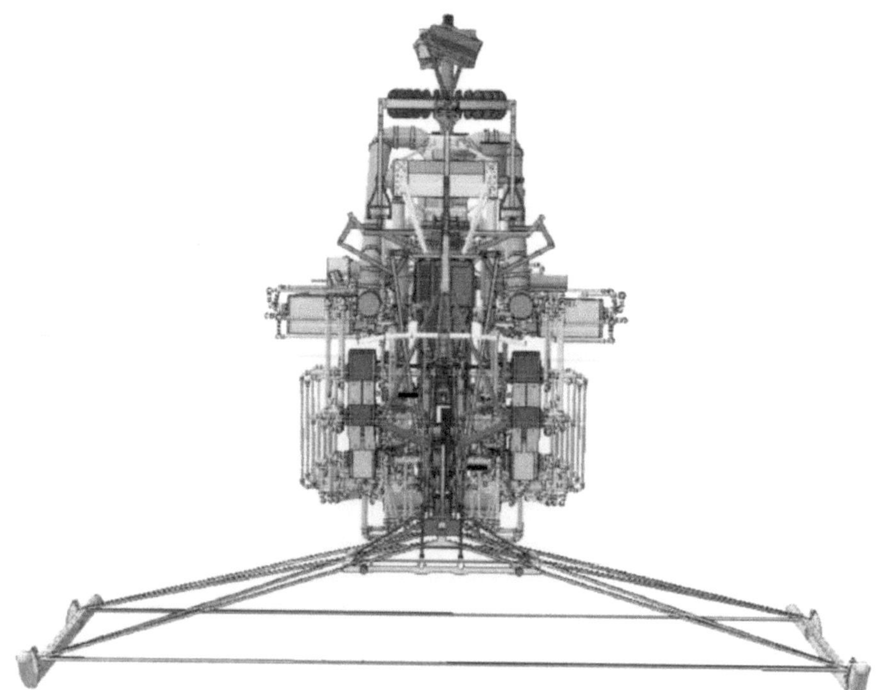

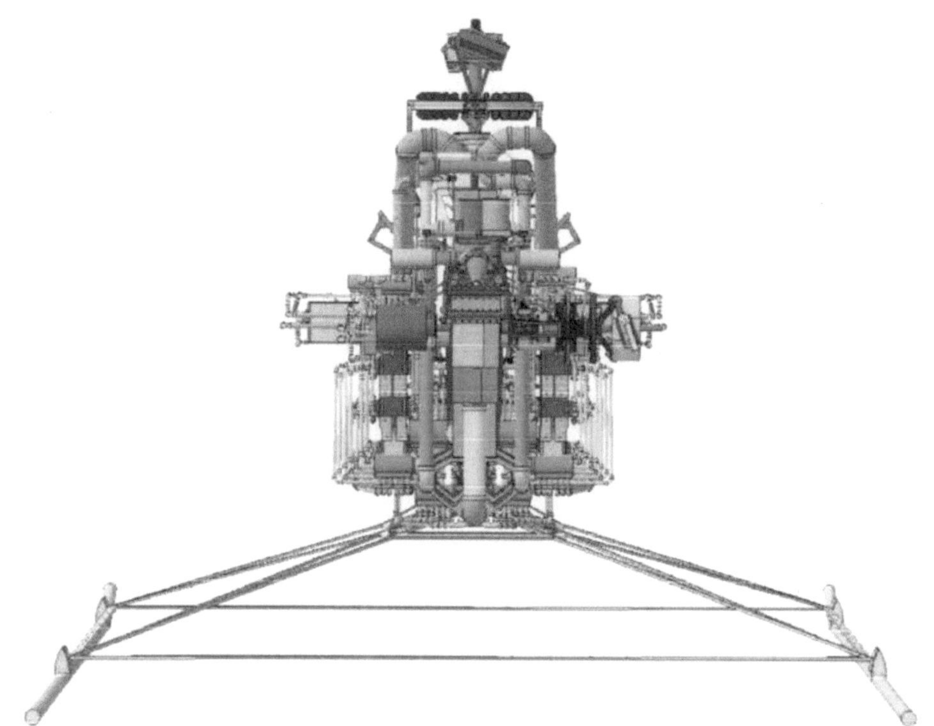

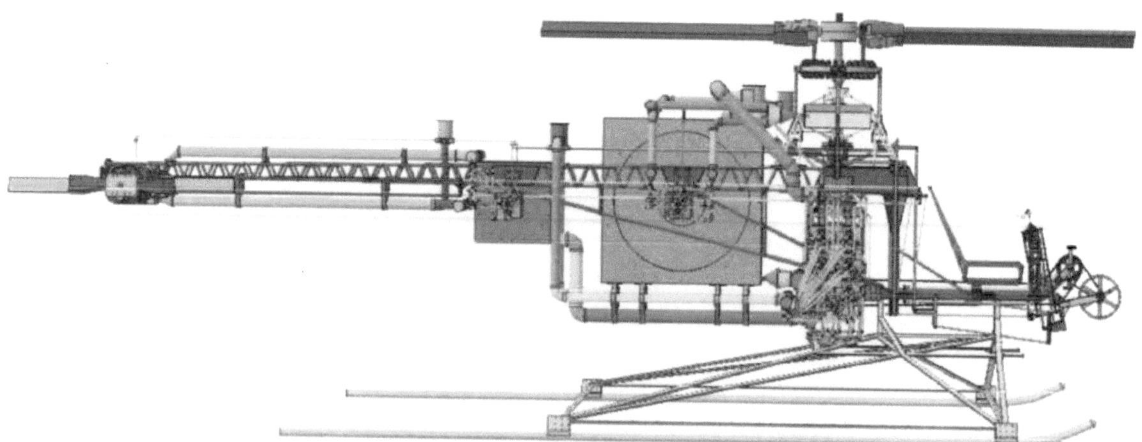

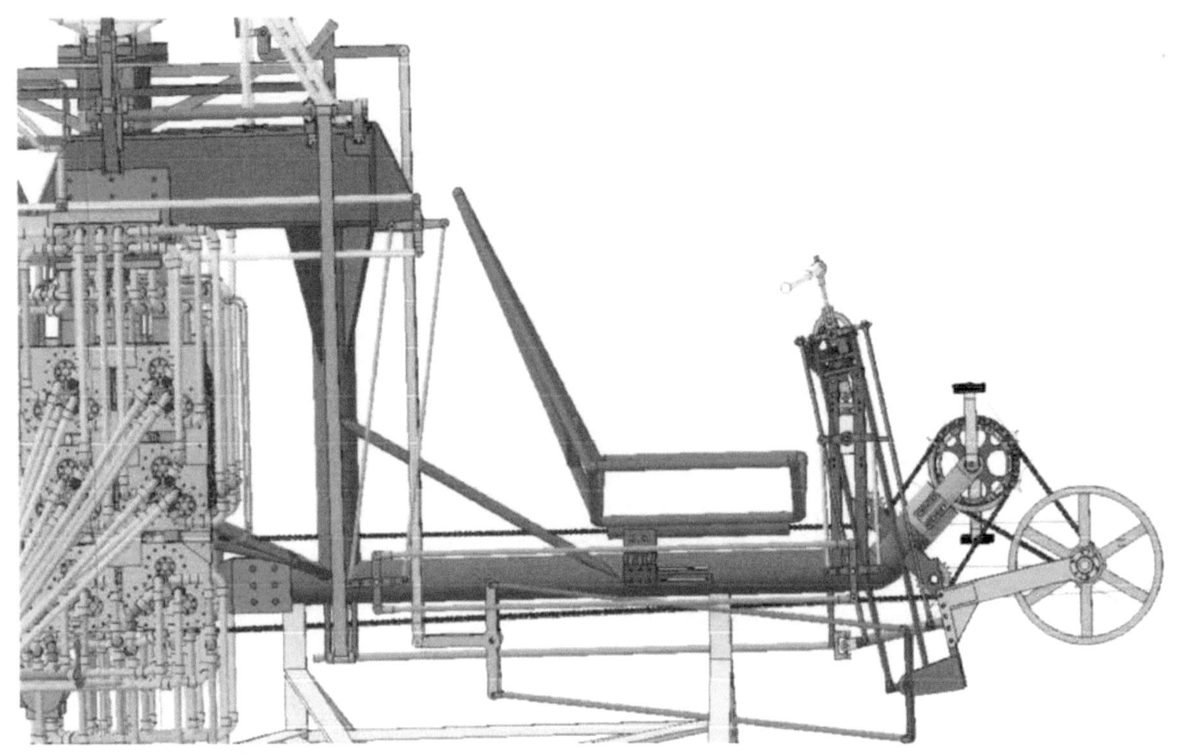

HPH3

Lift blade equations resulting in applied revolutions

$\frac{2Ph}{1.5} 2 = \text{area} \times$

$\frac{2 \cdot 3.518 \cdot .522}{1.5} \cdot 2 \times 79 \times 5 = 1,934,3371 \text{ 3in}$

$(91.775^2 \pi - 12.775^2 \pi) \cdot 2.394 = 65,684.04 \text{ 3in}$

$\frac{65,684.04 \text{ 3in}}{1,934,3371 \text{ 3in}} \; -2^n/2^n + n = 2^{5.0612}$

$\frac{3,500 \text{ lbs}}{.076521 \text{ lb/3ft} \cdot \frac{1,934,3371 \text{ 3in}}{12^3 \text{ 3in/3ft}} \cdot 5} \; -2^n/2^n + n$

$8173.1866 \; -2^n/2^n + n = 2^{12.9954} \text{ Rev/2c/R}$

$10.637^2 \pi \cdot .3125 - 10.325^2 \pi \cdot .3125 = 6.4208$ 3in

$\frac{3500 \text{ lb} \cdot 67.368 \text{ in}}{10.325 \text{ in}}$ (2500 lb) 22, 836.6102 lb at the 1.ft impeller blade vanes

16,311.8644 lb

$\frac{22{,}836.6102 \text{ lb}}{\frac{52.1 \text{ lb/3ft} \cdot 6.4208 \text{ 3in}}{12^3 \text{ 3in/3ft}}} - 2^n / 2^n + n =$

$117{,}963.[illegible]41 - 2^n/2^n + n = 2^{16.8}, 2^{16.2857}$

84,259.8099

$6.4208 \times 16.8 = 107.8693$ 3in $\times 12.99[illegible]$

1401.8047

$\frac{107.8693 \text{ 3in}}{\left(\frac{2.2262}{2}\right)^2 \pi - \left(\frac{1.8599}{2}\right)^2 \pi}$ (5465)

$\frac{5.14}{\pi} = 2.2282$

$\frac{16 \cdot .5}{\pi} = 2.5465$

$\frac{16 \cdot .5}{\frac{\pi}{2}} - (5.6866)\, 2 = 1.8599$

$= 6.1452$ in $\quad$ 2 PTUs / 6 mgT / 8 rev./2 crank C/R

~~7.6016~~ IN gears [illegible]

[illegible] ft impeller PTUs

22,996.6122 in

$\left(\frac{55}{2} + .01\text{ in}\right)^2 \pi - \left(\frac{2.5465}{2}\right)^2 \pi\Big) \times$ 6.1452 in = 3 in

12.3482

= .2463 in .4375 3 in

.5852 in

$\frac{15.5}{\pi} = 2.2273_D$

$\frac{17.5}{\pi} = 2.7056_D$

$\left(\frac{17.5}{\pi/2} - .5\right) \cdot 6.4612 = 2.019_D$

$\left(\left(\frac{2.7156}{2}\right)^2 \pi - \left(\frac{2.7056}{2}\right)^2 \pi\right) \cdot$ 12.3482 in = .5258 3 in

× 6.1452

× 3

= .7850 3 in + 3.1398

+ .2463 .5852

1.0313

.4375 × 2

+ .5258 × 6

1401.8017

2/6/8

6.122 × (numerator − 1.0313)

= 6.6122 new

secondary

PTU mat depth

$(\frac{2.5465}{2})^2 \pi - (\frac{1.8599}{2})^2 \pi \times$ ~~7.6016~~ 12.3482 $\times 2 =$ ~~36.1254~~ 58.6829 3in

$(\frac{2.7056}{2})^2 \pi - (\frac{2.019}{2})^2 \pi \times$ ~~7.6016~~ 12.3482 $\times 6 =$ ~~116.2024~~ 188.7616

116.2024 + 36.1254 = 152.3278 3in
247.4445 3in

~~116.~~ 152.3278 3in (247.4445) × 7.6016 in (12.3482) = 7.7553 in new impeller

152.3278 3in (247.4445) − 3.0192 3in (4.0296)

PTU gears depth 12.5526 in

22,836.6102 lb × 2.5465/2 in

~~2.5465 in~~
~~2.5465/2 in~~

= 692.3027 lb

42 in primary impeller radius

23000 lb = 60,000 lb/2in · πr^2

23000 / 60000 · π = .347 in.

46000 / 60000 · π = .494 r

.494
.493

52 T. @ .375 Dp [illegible] 7" length results in 110 lb at the crank operator force at full power output. Easy, and 3500 lbs of LA applied force

20° 851.3722 lb drag

2649 lb lift

3000 lb · sine 20° · [illegible] = 904.8[illegible]

964.1814 [illegible]

2800
2700
2650
2649

851.3722
851.6936
867.7633
899.9027
964.1814

[illegible] 11.08:

Empty manifest

$\frac{.417 \cdot 2.020 \cdot 2}{1.5} \times 11 \times 5 = 61.7716\ ^{3}\text{IN}$

$17.779^2 \pi \cdot 1.378 - 6.779^2 \pi \cdot 1.378 = 1{,}169.4584\ ^{3}\text{IN}$

$\frac{1169.4584\ ^{3}\text{IN}}{61.7716\ ^{3}\text{IN}} \; {}^{n}/_{2^n} + n = 2^{4.1832}$

$\frac{2836.6102 \times 67.368''}{144''} = 10{,}683.7275\text{ lb} \times \sin 20° \times \cos 20°$

$= 14{,}117.4113\text{ lb}$

$\frac{14{,}117.4113\text{ lb}}{\frac{.07651\ \text{lb}/_{3\text{ft}} \cdot 1{,}169.4584\ ^{3}\text{in}}{12^3\ ^{3}\text{in}/_{3\text{ft}}}} \; {}^{2n}/_{2^n} + n = \frac{2^{18.0101}}{2^{4.1832}} =$

$2^{18.0101} \times 2^{4.1832} = 22.2233$

Revolutions per 2 crank [illegible] at full throttle.

$\frac{\text{aerodynamic force to tail rotor blade vanes applied at 12 ft from lift blade axis.}}{\text{air density} \cdot \text{swept volume of tail rotor blade vanes} / 12^3\ ^{3}\text{in}/_{3\text{ft}}}$

${}^{2n}/_{2^n} + n =$ [illegible] multiply [illegible] rotor blade vanes [illegible] tail rotor blade [illegible] swept volume 1 [illegible].

$$\frac{14117.4113\text{ lb} \times 13.465\text{ in}}{\frac{10.875\text{ in}}{2}} = 34{,}180.3623\text{ lb}$$

$$\left(\left(\frac{11.5}{2}\right)^2\pi - \left(\frac{10.875}{2}\right)^2\pi\right).5 = 5.4917\ ^3\text{in}$$ final impeller gallery volume 1 C/R.

$$\frac{34{,}180.3623\text{ lb}}{\frac{52.1\text{ lb/ft}^3 \cdot 5.4917\ ^3\text{in}}{12^3\ ^3\text{in/ft}^3}} - 2^n/2^n + n = 2^{17.5749}$$ staticfold force multiplier

$$12 \cdot 5.4917 \cdot 17.5749 \times 22.2233 = 2{,}144.6324\ ^3\text{in}$$

$$\frac{2.5465}{2}{}^2\pi - \left(\frac{1.8599}{2}\right)^2\pi$$

$$6.2678\left(\left(\frac{2.5565}{2}\right)^2\pi - \left(\frac{2.5465}{2}\right)^2\pi\right) \cdot .2512 = .5024\ ^3\text{in}$$ 2 PTU

$$+\ 6.2678\left(\left(\frac{2.7156}{2}\right)^2\pi - \left(\frac{2.7056}{2}\right)^2\pi\right) \cdot 3 = 1.6012 \quad .8006$$

6 mgt / 12 Rev / 2 crank C/R

$$\frac{2144.6324}{\frac{2}{6}\ 12}$$

14.8933 ^3in per PTU 4.1022 ^3in 1.0518 2.1036 ^3in

$$\frac{2144.6324}{(14.8933 \cdot 1.1056)} = 12.598 = 6.2678$$ inches initial gears depth for tail rotor secondary PTUs; each side.

$\times\, 6.2678 =$ 6.7442 in new gears depth $=$ [illegible] final gears depth

$\left(\left(\frac{84.125}{2}\right)^2\pi - \left(\frac{83.5}{2}\right)^2\pi\right).5 = 41.1414\ in^3$ primary impeller vol. C/R

$\frac{692.3027\ lb}{}$

$52.1_{lb/ft^3} \cdot 41.1414\ in^3$

$\frac{}{12^3}$

9.0901

$41.1414 \cdot 9.0901 \cdot 8 = 2,991.\ldots\ in^3$

impeller blade vane gallery C/R · applied static load · Rev's

$\left(\left(\frac{2.5465}{2}\right)^2\pi - \left(\frac{1.8599}{2}\right)^2\pi\right)$

$= 6.5578$ Primary PTU gear depth, initial.

8 PTU / 24 inst

$\left(\left(\frac{2.2281}{2}\right)^2\pi - \left(\frac{1.8599}{2}\right)^2\pi\right)$

.2628

$\left(\left(\frac{2.5565}{2}\right)^2\pi - \left(\frac{2.5465}{2}\right)^2\pi\right) 8 = .2806$

8/24

15.5824

$6.5578\left(\left(\frac{2.7156}{2}\right)^2\pi - \left(\frac{2.7056}{2}\right)^2\pi\right)\frac{3}{24} =$ .9012 .8377(15.5824 − 8.8038)

5578

15.0748 2W gear depth

6.9304

tail rotor primary impeller applied tail rotor secondary output PTO:

34,180.3623 lbs

23

$1637.4167 \times \text{Sine}\,20 \times \text{Cos}\,20 + 1637.4167 = \frac{2163.6723\ \text{lb} \times 13.165\ \text{in}}{10.875\ \text{in}} = 2619.2869\ \text{lb}$

$\frac{2619.2869\ \text{lb}}{.07651 \cdot \frac{1{,}169.4584}{12^3}} - 2^n/2^n + n = 2^{15.5437}$

$\frac{1{,}169.4584\ 3\text{in}}{61.7716\ 3\text{in}} - 2^n/2^n + n = 2^{4.1832} = 2^0$

$2^{19.7269}$ Rev.

$\frac{2619.2869\ \text{lb}}{\frac{52.1\ \text{lb}/\text{ft} \cdot 5.4917\ 3\text{in}}{12^3\ 3\text{in}/3\text{ft}}} - 2^n/2^n + n = 2^{13.931}$ static (solid fuel)

$5.4917 \times 13.931 \times 19.7269 = \frac{1{,}509.2088\ 3\text{in}}{\left(\frac{2.5465}{2}\right)^2 \pi - (\ldots)} = 10.585\%$ initial gears depth

2 PTO / 6 mgt points / 5 Rev / 2 crankshaft

$10.5857\left(\left(\frac{[illegible]}{2}\right)^2\pi - \left(\frac{[illegible]}{2}\right)^2\pi\right) = .4243 \text{ 3in}$

$10.5857\left(\left(\frac{2.7156}{2}\right)^2\pi - \left(\frac{2.7056}{2}\right)^2\pi\right)3 = +1.3522 \text{ 3in}$

1.7765 3in clearance volume in 1 PTU.

$\frac{1509.2088 \text{ 3N}}{2/6/5} = \frac{25.1535}{(25.1535 - 1.7765)} \times 10.5857 = 11.3901 \text{ in}$ final tail rotor secondary PTU gears depth.

$\frac{2629.2869 \text{ lb} \times \left(\frac{2.5465}{2}\right)}{24 \text{ in}} = \frac{138.5586 \text{ lb}\left(\frac{2.5465}{2}\right)}{\frac{54.375}{\frac{\pi}{2}} - .375 \cdot .6866} = 2.9654 r$

$\frac{138.5586\left(\frac{2.5465}{2}\right)}{2.9654 r} = 59.4927$

results at about 110lbs + 59.4927lbs ≊ 170lbs at the crank

$\frac{170}{258} \times 100 = 65.8915\%$ of approximate body weight force at the crank

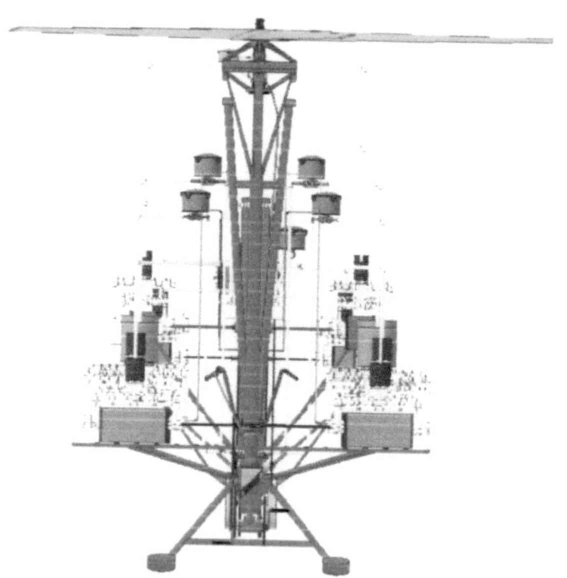

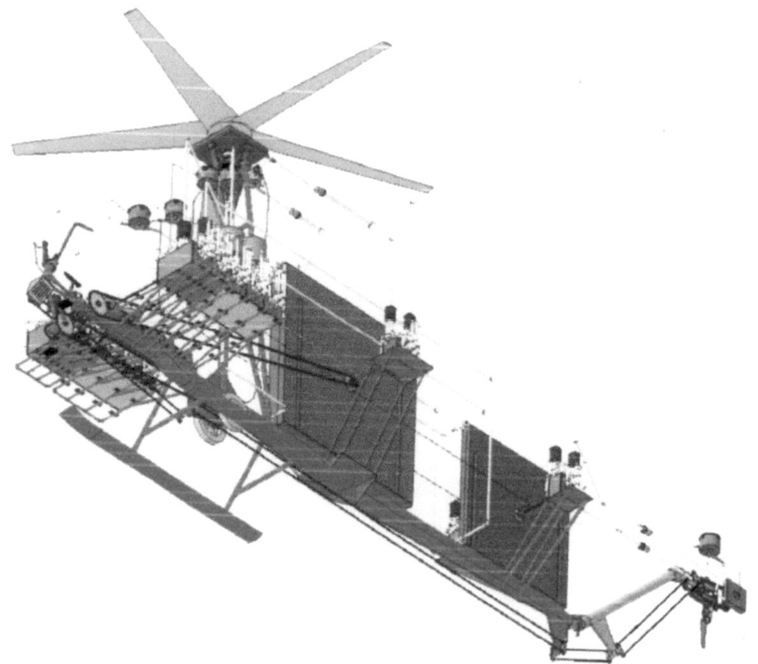

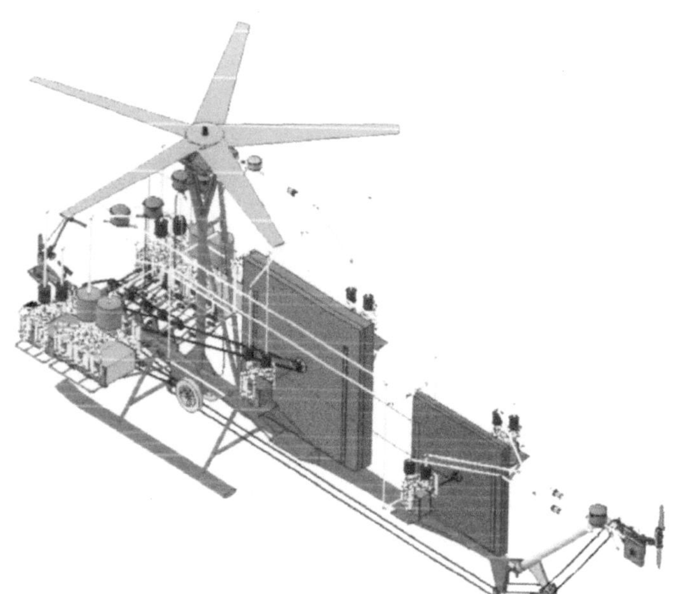

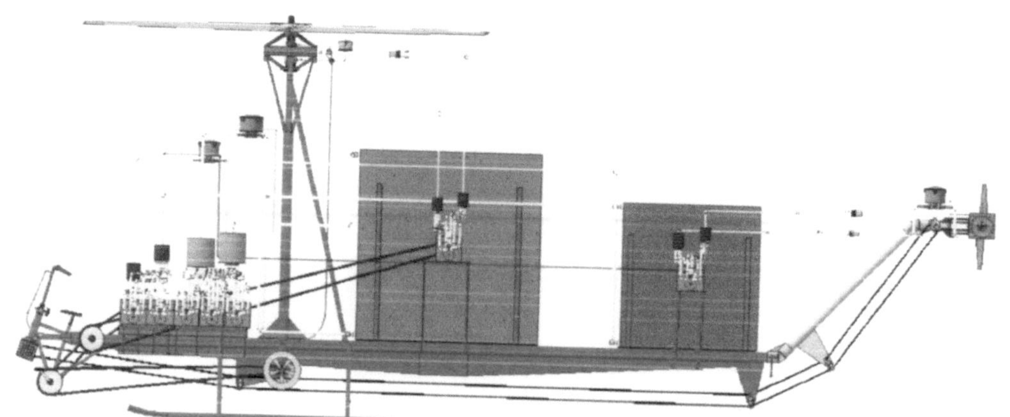

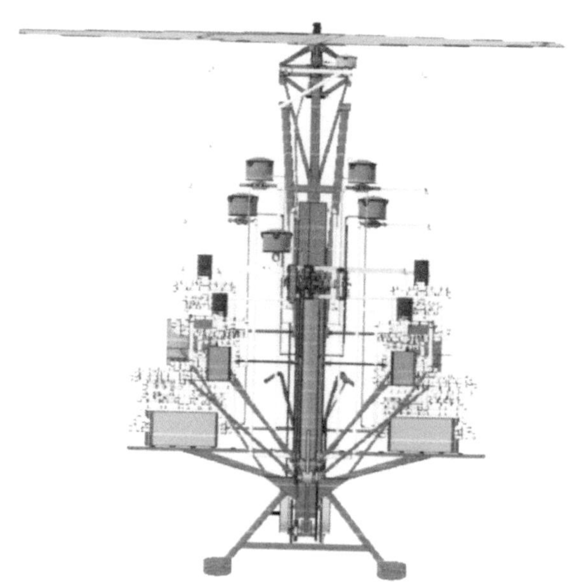

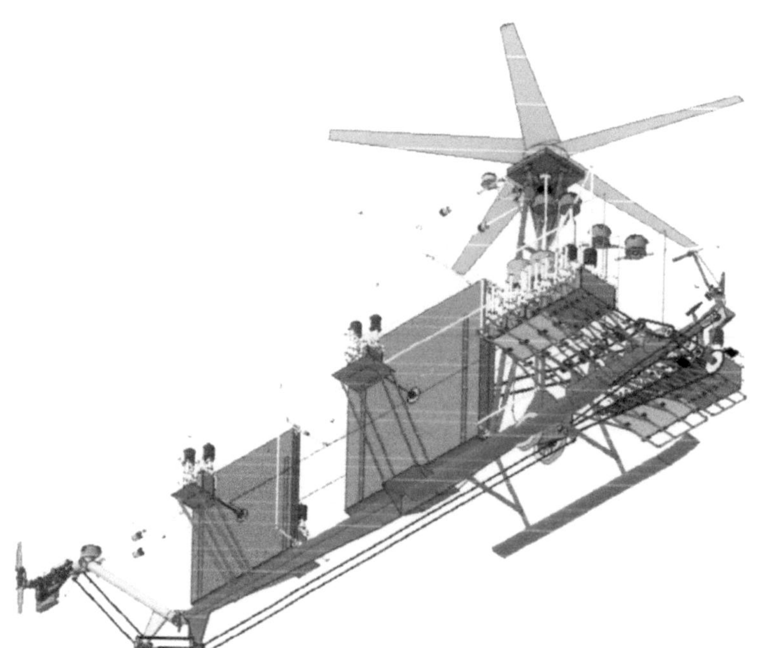

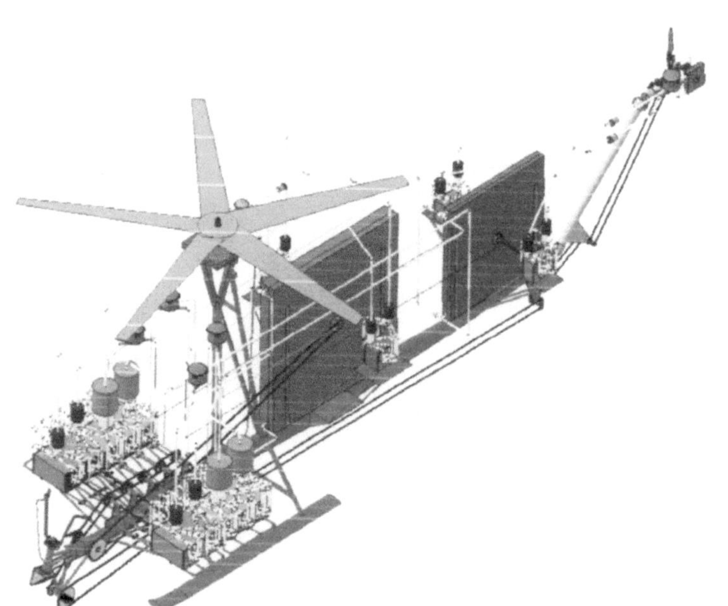

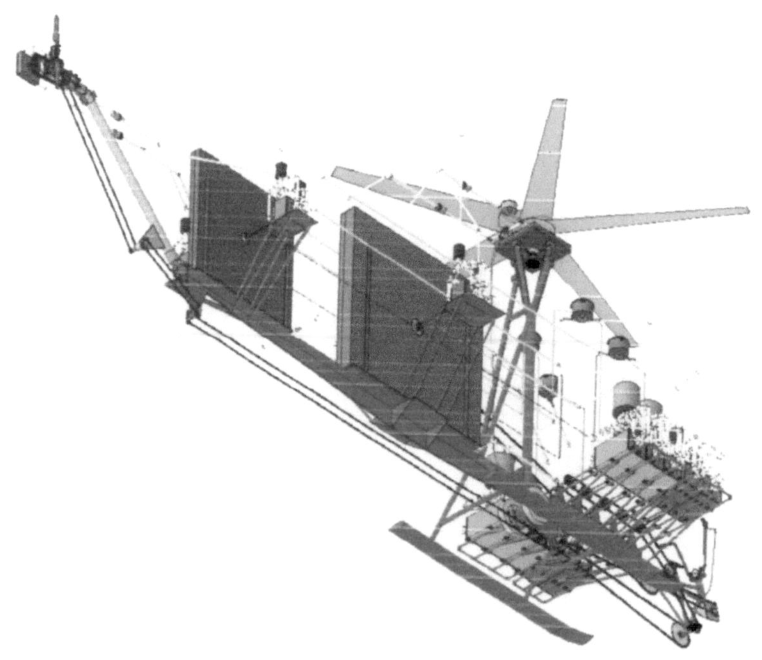

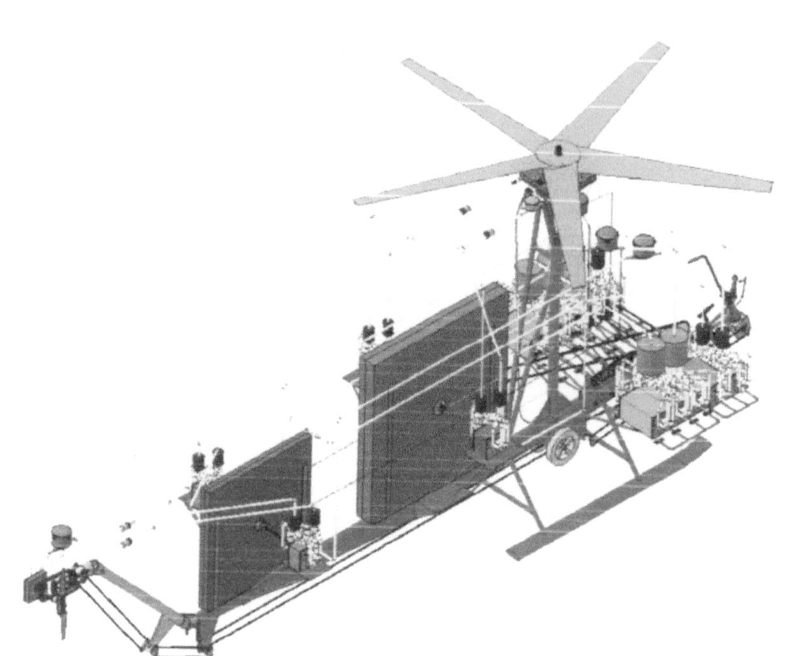

Bicycle at an Optimum Performance for Velocity

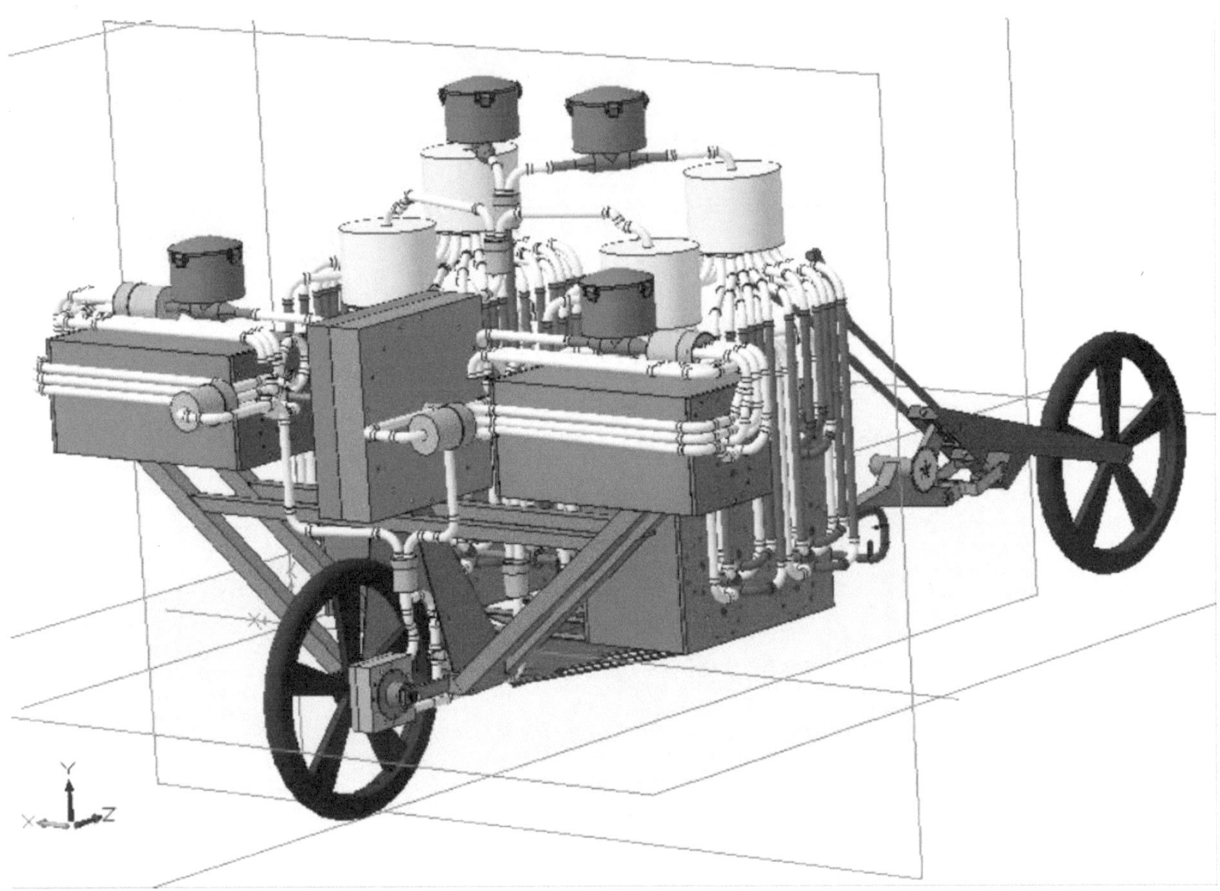
Y
X
Z

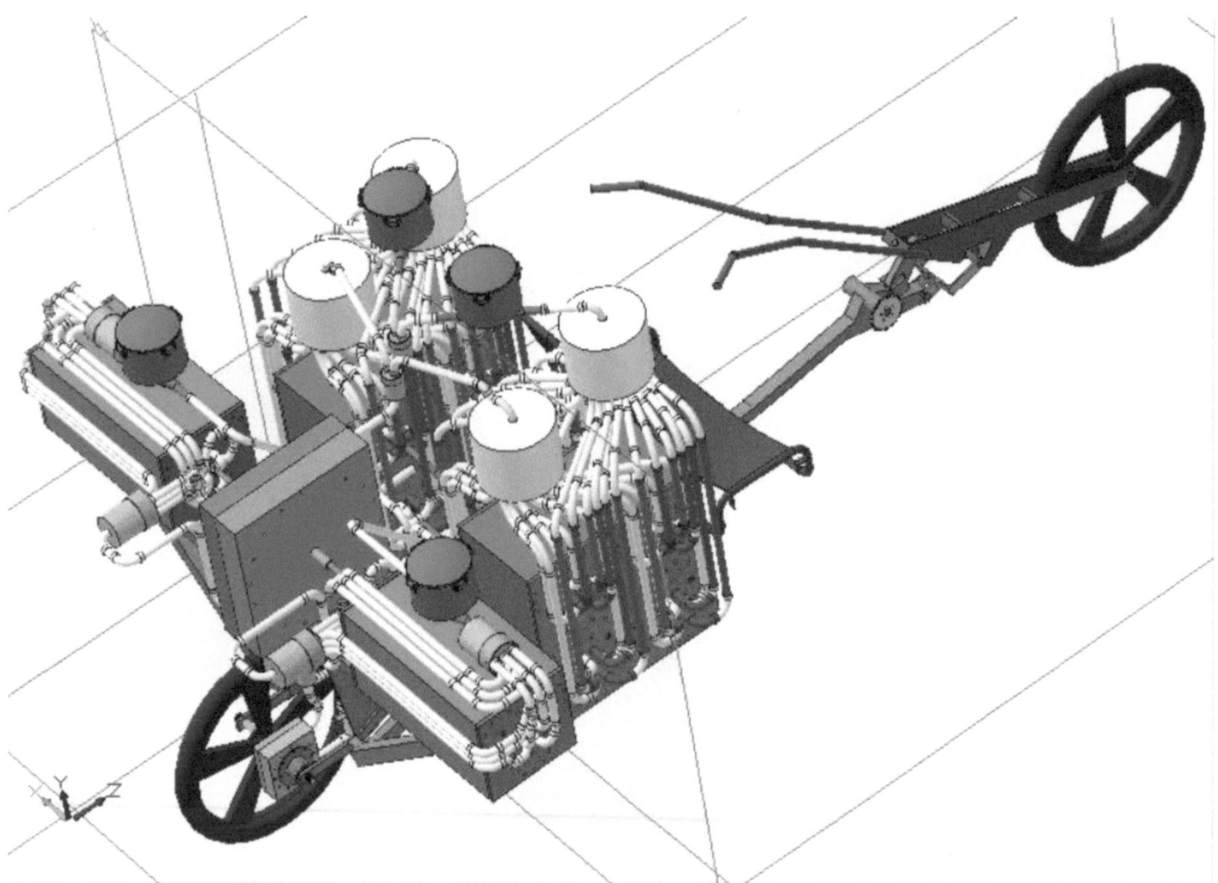

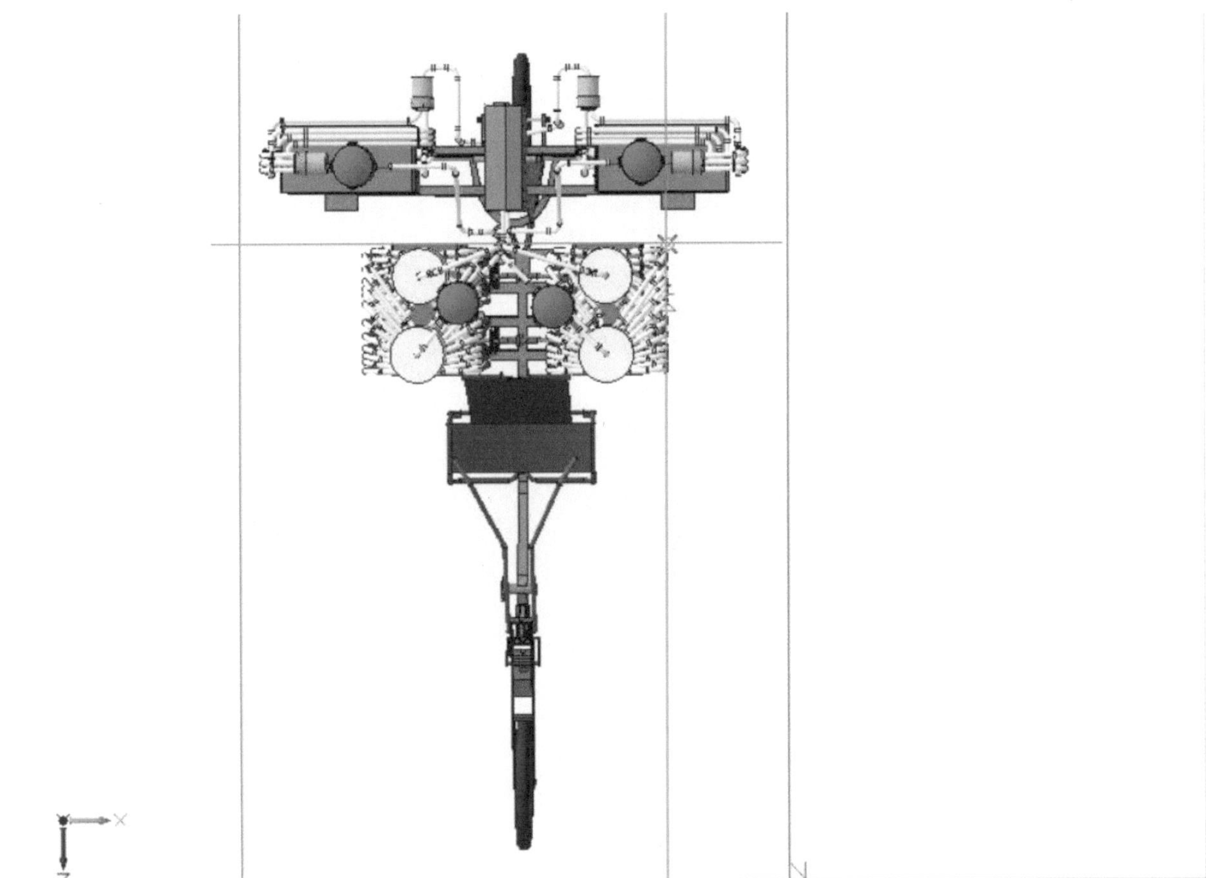

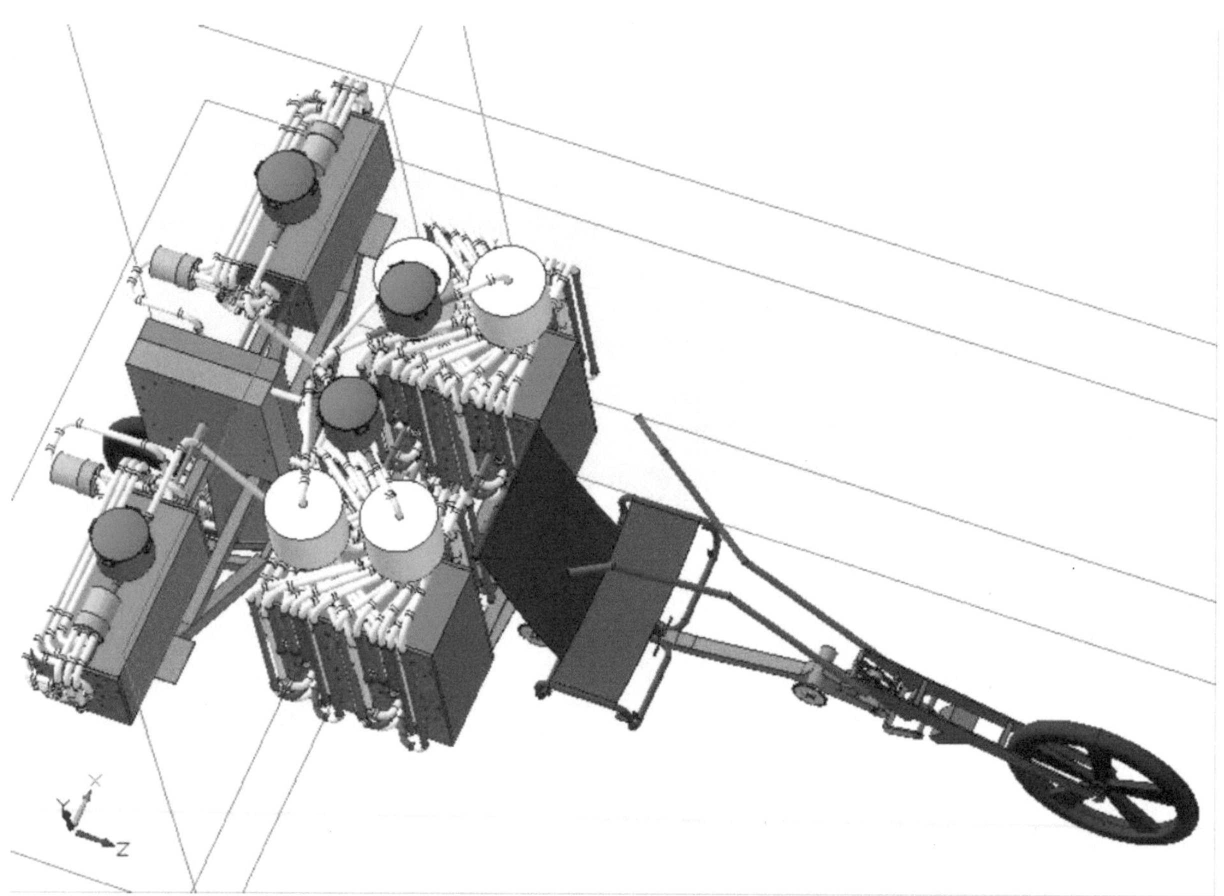
X
Y
Z

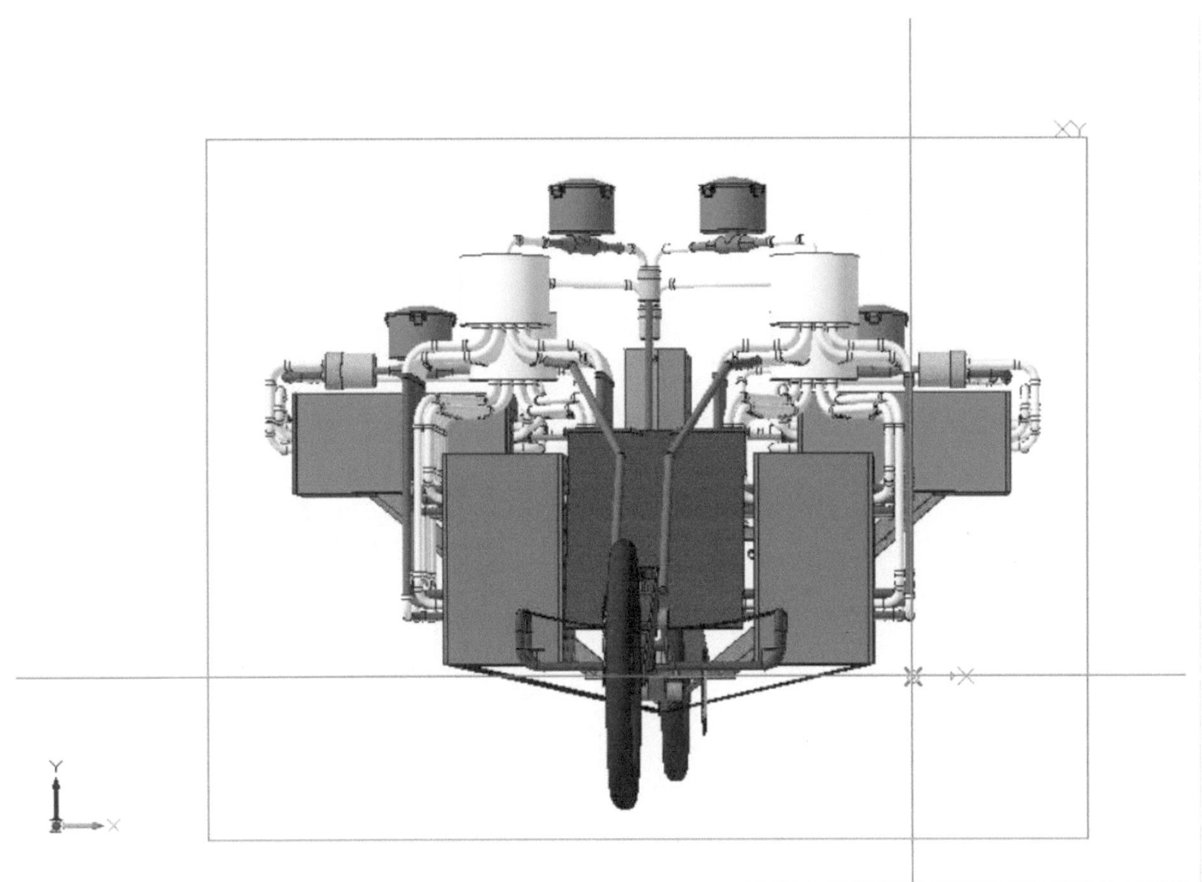
XY
Y
X

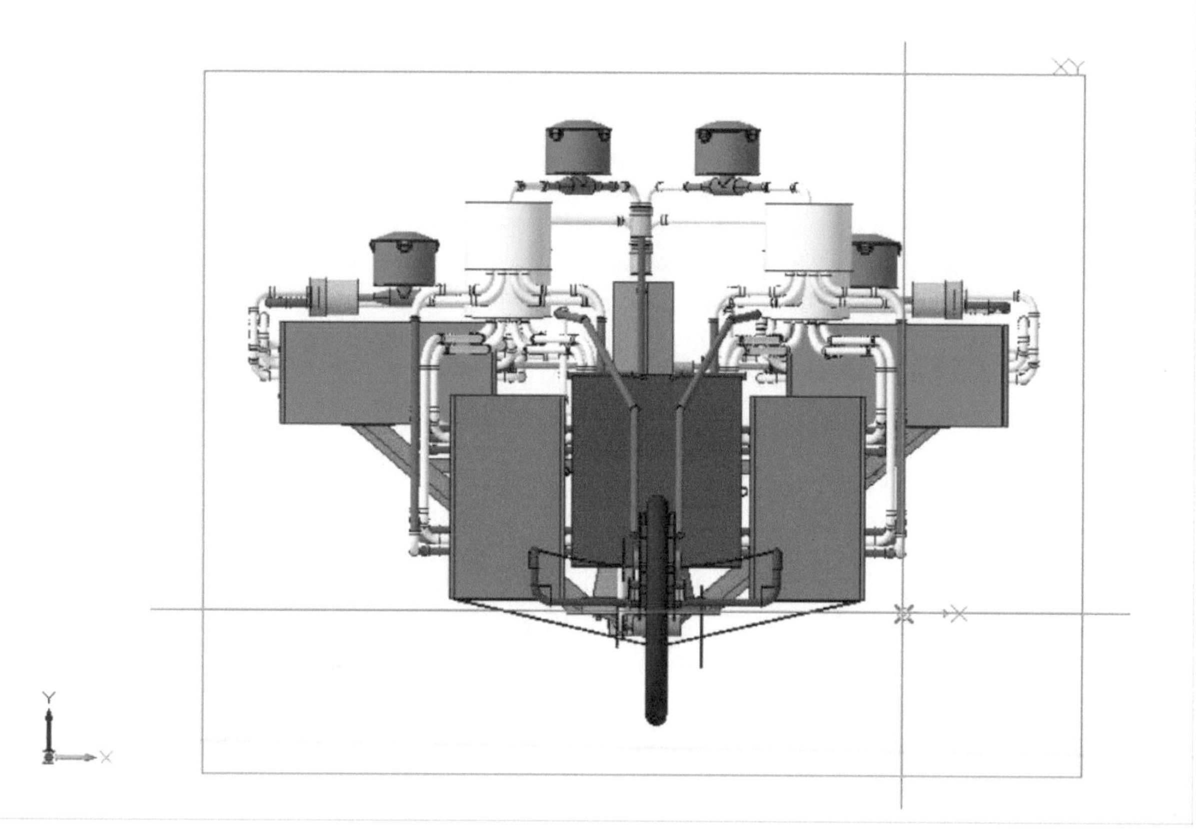
XY
X
Y
X

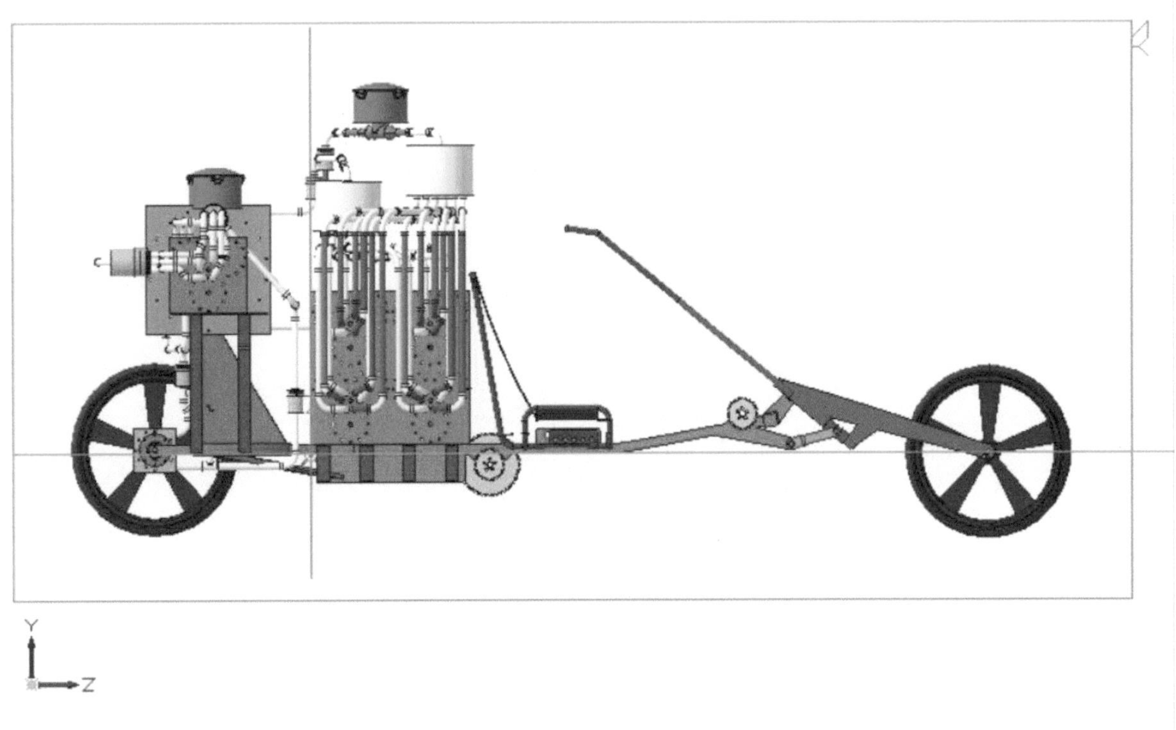
Y
Z

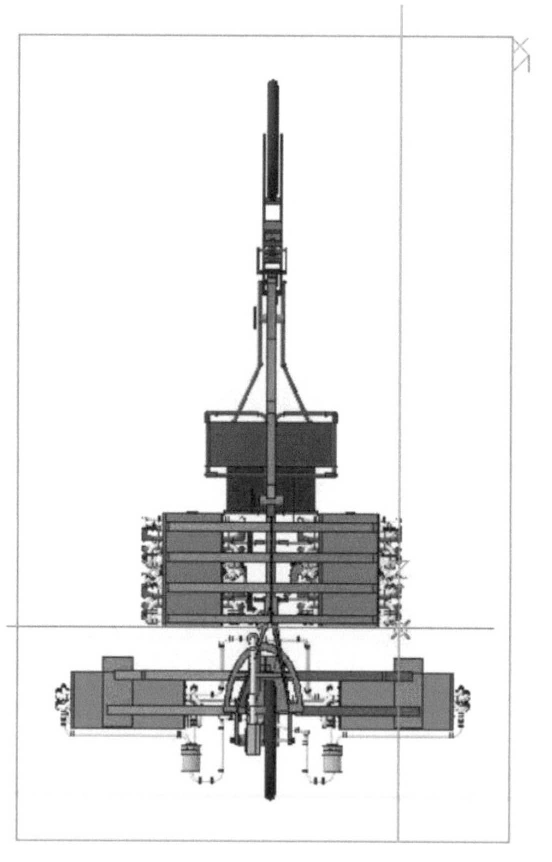

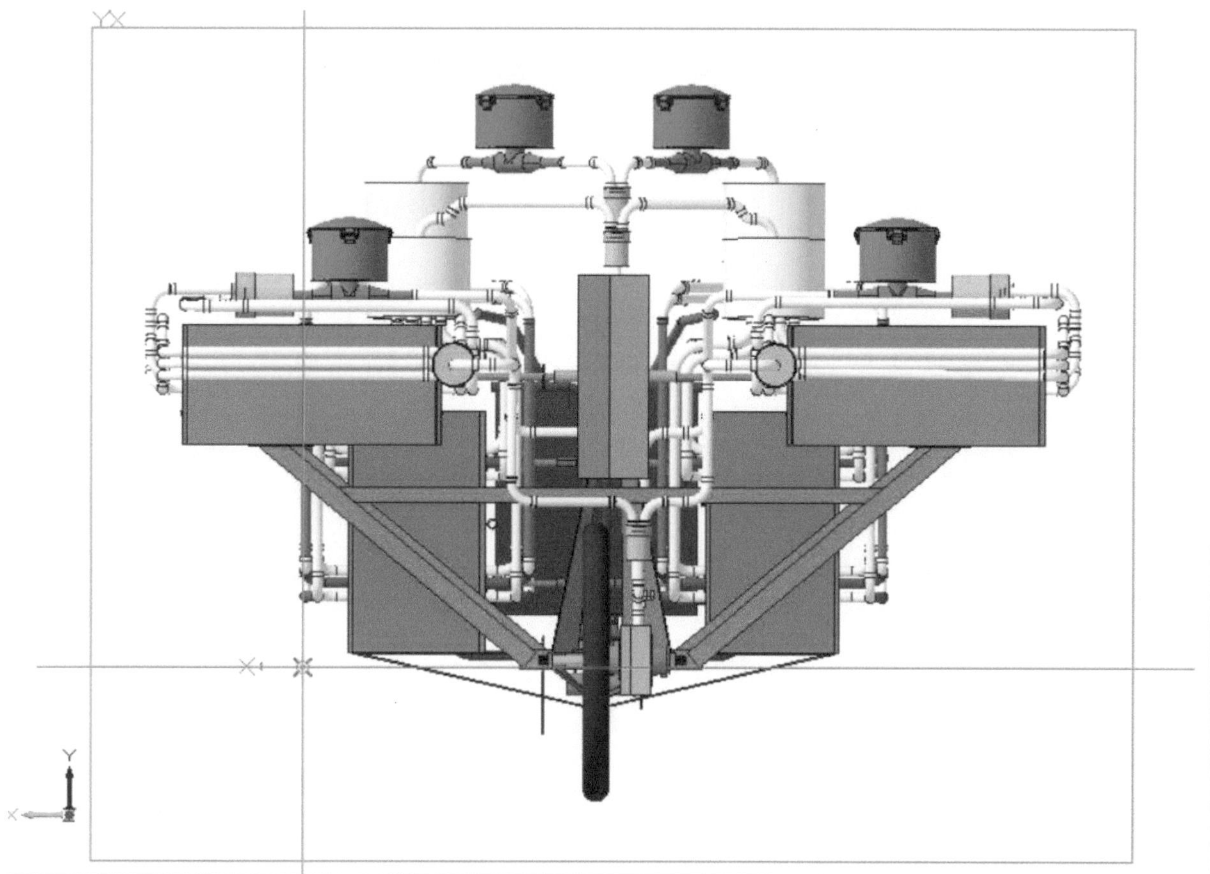
Y
X

Self-propelled Motor

Somehow find the monetary resources to begin:

In the meantime be patient, continuing in the faith, be diligent, win the race. All things work together for good to those who love God.

First design a self-propelled motor to fit in to an engine compartment for a vehicle without an engine. The Prototype design is negligible. Find a vehicle online without an engine ($$$). Take engine compartment dimensions for bolt holes, shaft input to transmission etc, and design the motor to be accepted in to the engine compartment.

[Machine shop doesn't know how to design one or explain how it works: not your subservient, afraid of the repercussions of freedom, provide myself with a defense continuously, keep my guard up all the time.]

Isaac has to work first, take camcorder to storage compartment and film first attempt and assembly, make DVD, include with application for patent pending in the mail. Figure out why not if it doesn't work and fix it.

Go through lba.org (Louisville Bar Association) to find a patent attorney, contact David Carrithers.

Apply for patent, and get a patent pending ($$$),

{Send a professional letter snail-mail to the Louisville Science Center with a DVD of the motor offering to lend it to them for exhibit (answer: Y/N.). Y: Loan Isaac to the Louisville Science Center for public exhibit, and wait for an offer for a dynamometer test. N: Continue working.}

Continue working to compete in the X-Prize without including the media,

---

If Isaac proves, the order of enrollment should be according to history in kind, as per the capability of manufacturing since all the employable applications may be taken in to account all at once:

Agriculture: Combine, Tractor, Harvester, other major farming equipment, etc.

Electric Power: City power generating plants,

Commerce: Rail, Maritime, Tractor Trailer, Short delivery local,

Military: As provided determining on the result of the output of Isaac,

Experimental Research: Mining, Drilling, Exploration, Reclamation, Refining, Chemical research, all aspects involving analytical research in the sciences,

Infrastructure: heavy construction equipment, road maintenance, earth moving equipment, etc,

Fire: Fire Truck, fire equipment with power-plant operation, anything which can be used by a firefighter that is powered by a motor,

Justice: Any Police Vehicle or Police defensive mechanism operating on a power-plant, etc,

Health and Human Services: Ambulance, Helicopter, etc,

Education: School bus, School should be powered by local power-plant,

Interior Department: Wildlife and Forestry transportation and power driven equipment, ATV, off road vehicle motorcycle/automobile, fire equipment, earth moving equipment, heavy construction equipment,

Personal Transportation Means: Cars, trucks, motorcycles, scooters, ATVs, boats, weed eaters, lawn mowers, etc.

Space: Locomotive creation of electric power to electrify space transportation systems, transference of atmospheric pollutants in to clean/breathable air while in space and on the Earth, generation of electric power for defense in Space and for the nation, Cognitive use of the generation of power to reduce hunger in the world and throughout Space,

Recreation: boats, motorcycles, cars, motorhomes, scooters, ATVs, snowmobiles, personal watercraft, anything a person can buy at a store that they can put in their garage for personal use in application with a gasoline or diesel powered power-plant.

Sports: automobile racing, hydrofoil racing, go-karts, World-Cup Sailing electric power generation,

How to Start My Own Business

List of Essential Machine Tools and Equipment

Coolant, $27.95

Spray Mist Coolant System, $199.99

10 Gallon Flood-type Coolant System, $149.00

Anti-fatigue Matting, $55.00

Adjustable Wrench Set, $48.95

7 pc. Roll Pin Punch Set, $39.95

Locking Plier Set, $45.95

Die Grinders, $55.00

Die Grinder Kit, $67.00

Vertical Bar Storage Rack, $299.00

Tap Magic Cutting Fluid 24 Ct. Case, $56.40

Soluble Oil 5 Gal., $122.99

Lathe, approx. $2995.95

Mill/Drill, approx. $2355.00

Horizontal Bandsaw, approx. $2179.00

Replacement Blades, approx. $19.99 - $45.99 each

Combination Belt/Disc Sander, $419.00

Carbide Wheel Tool Grinder, $1195.95

52 Pc. Combination Steel Step Block & Clamp Set, $65.95 - $79.95, or

59 Pc. Tooling Package, $199.95

Machine Vise, $119.00 - $1495.00

Granite Surface Plate, $32.95 - $569.95

Dividing Head (Super Spacer), $995.00

Quick Change Tool Post Set, $185.95 - $425.95

13 Pc. Collet Set & Tool Tray, $65.95

Digital Read-out System, $999.00 - $2349.00

Miscellaneous Carbide Double End Mill Bits, $9.99 - $59.97

Rounding End Mill Set, $125.95

5 Pc. Fly Cutter Tool Set, $25.95

Countersink Set, $85.95

Jobber Drill Set, $79.95

Reamer Set, $95.95

38 Pc. Tab & Die Round Set, $112.00, or

Hand Taps & Set, $15.75 - $23.74

Indexable End Mill, $45.95

Boring Bar Set, $55.00 + $67.45 + 77.50

Boring Head, $195.72

Carbide Inserts, $4.23 - $8.61 each

Turning Toolholder Ket, $139.00

Jeweled Test Indicator Set, $36.95

Dial Indicator (.0001 Grad.), $24.95

6 Inch Long Precision Parallel Set, $41.95 - $89.95

Outside Micrometer Set, $32.89

Dial Height Gage, $59.95

Electronic Depth Gage, $99.00

Edge Finders, $11.95 - $15.95

Forced Locking Magnetic Base, $24.95 - $47.95

6 Pc. Telescoping Hole Gage Set, $17.95

General Purpose Files, $17.79

Round File (Second), $5.39 each

Shop Roll Assortment Pack (Aluminum Oxide), $29.95 - $55.95

Hand Wire Brush (2x), $3.39 + $4.72

Twist-a-Burr Set & Replacement Blades, $8.95 - $12.95

Pedestal Fan, $105.00 (optional)

Multipurpose Welder Pack, $3769.00 - $8215 (Lincoln-Electric)

Filler Metal and Welding Rods

Plumbers Tape

Argon Pressure Vessels

Argon Pressure Vessel Refills

Any Additional Oversights of Required Equipment Not Listed Here

Estimated Total so far, on the low side: $23,201.00 for machine tools and equipment, not including real property zoned for business plus a prefabricated metal building, its construction, and the necessary utilities maintenance (electricity, sanitation/recycling, water, sewer,...) for the duration until the new business is self-sustaining

New Private Enterprise Funding Transactions Accountability

There will be a paper trail of all these financial affairs:

A ledger
Receipts for purchases made
Invoices for outsourced jobs
A Business Bank Account
Minimum Balance after all purchases are made for required items
A Business Bank Account Statement
Funding deposits may be made only to the business bank account
The sum of the sponsor's deposits will equal the cost of the items purchased on the bank account statement according to the formula given in the presentation.
There will be no additional balance accountability in the business bank account for items purchased with sponsors funding.
If there is any additional balance to the business bank account the author may have the deposit accountability for it.
Interest on the business bank account belongs to the new private enterprise.
Extraneous expenditures from any additional balance accountability may be accounted for. Elementary items, if any, with additional deposit accountability, e.g. personal hygiene, first-aid, creature comforts, luxuries, et cetera could exhaust the additional balance.
Sales and Use Tax, City, and Federal Tax Accountability: all funding is taxable.

Prospects

Due to the possible popularity of the available items to be for sale by the new private enterprise it may be unlikely that the new private enterprise will struggle to stay in business. If the self-propelled motor works and proves to be a reasonable source of power and locomotion it may be likely to replace the internal combustion engine and all its capabilities. Nevertheless, the physically powered transportation means should not fail and a variety of transportation means are likely to come from the capability of the manually powered types of transportation which may be engineered.

The manually operated primary and secondary power-plant for the human powered transportation means are constructed from tested and proven mechanisms that have been in use for over 100 years, they are just designed in a different configuration (constructive reduction to practice). The author of this presentation also would like an opportunity to compete at the Salt Lake in Utah to attempt to exceed the sound barrier in a human powered vehicle if it can be designed.

Purchase Guarantee

Commitment to purchase all of the real property zoned for business upon which is the commercial building of the new private enterprise is existing, the commercial building, and the elements of proprietary business within it at the time of obtaining financial capability is guaranteed, providing the cost is the remainder owed if not the full purchase cost, at the present value for the real property and the commercial building and the present value for the elements of proprietary business within the commercial building and any on the real property as guaranteed by an independent real property appraiser, or the difference at which is the going rate for the real property and commercial building value and elements of proprietary business within the commercial building and any on the real property at the present time depending on whether the amount has been paid down by payments.

No choice is granted on the matter of ownership of the real property and commercial building and elements of proprietary business within the new private enterprise to the sponsors when all ties to the sponsors are severed, the author will own the new private enterprise, and at this time the new free enterprise has not gone public and the author will have preemptive right.

This agreement may be signed, dated, and notarized, and denoted below as may be required for legal purposes. Anticipation by contract made by this business may be considered harassment by the author of this work. A affidavit between this company and the author for any verbal transaction between the author and this company at all must be secured by an original affidavit provided by the author, signed by both parties, dated, and notarized in the presence of both parties, and a copy made to be in the possession of this company.

Loose Employment Policy for This Business:

Since you will not be being paid to work at the beginning of your on-duty hours, I duly inform you that demand for the units produced by this business may or may not result in a high-pressure environment once start-up is successful and the business can sustain itself on its own income from retail sales of its own products. The goal will be to maintain production at the limit of what the demand will be once the back-orders have all been filled, assuming there will be a demand. One unit produced by this company has a lifetime transferable warranty, and as such, due diligence is necessary for parts production for that unit:

1. Do not quit your present job to come to work here.
2. On duty hours here are free-time-at-your-leisure hours after you get off work from your regular job.
3. You are expected, at this business, to do something pertinent and relevant with respect to the projects at hand, gaining experience on the job.

4. There are no restrictions on when you have to work, except that you do not quit your regular job to work here, and you come to work here when you are not working at your regular job.
5. You may come and go as you please
6. Please clock in when you come to work, and clock out when you leave or take an extended break.
7. If you start working at this business unemployed you may work up to 10 hours per 24 hour period, for your own safety, The number of hours you worked at your regular job subtracted from 16 hours = the number of hours you may work at this business per 24 hour period if it is not more than 10 hours, for your own safety.
8. You do not have to put in a 40 hour work week.
9. Any free time put in as clocked-in work over 40 hours will also be clocked in to receive back-pay for the overtime
10. You do not have to show up for work
11. If, within a 24 hour period, you have clocked-in (no on-the-job time limit required) and, you are tired and need to rest you may; you may even clock-out and leave the premises and come back inside of 24 hours and clock in up to 10 hours on-the-job in a 24 hour period as per rule 7.
12. Using the necessary facilities is not considered an extended break.
13. Extended breaks, including meals, need to be clocked-out for and clocked-in from for accurate record-keeping.

14. You may take time off without coming to work if you need time off, without calling in; you do not have to come back to work if you don't want to. You must be clocked out.
15. After an extended leave of more than 90 days clocked-out you will be considered absent
16. When the business begins functioning on cash flow from income generated from sales and it is possible for you to receive a steady wage you will begin to receive back-pay and your hours on-the-job may or may not become more regular and performance demand may or may not increase.
17. The business should be open 24 hours once the start-up situation has settled in.
18. Holidays are Christmas Eve, Christmas Day, and New Year's Day
19. If, after the business has successfully started up, your attendance or contributions per month or six months or year is/are not significant enough to warrant your participation on the job you may be asked not to come back to work.
20. This Loose Employment Policy for This Business may or may not change at the point of successful start-up.

Type 1 human powered Aircraft Warranty

The purchased aircraft, with the exception of the pontoons, hydraulic fluid, grease, chain lube oil, chain, alternator, battery, electrical wiring and electronic instruments, is warrantied against defects in manufacturing and workmanship for the life of the product with these exceptions: Normal wear-and-tear causes some parts to become defective over time: parts that wear over time e.g. drive and driven sprockets, bearings, grease seals, seat fabric, handlebar grips or handlebar tape, pedals; are not covered from normal wear and tear by the warranty. The all-weather cover is not covered by the warranty. Any physical injury or death resulting from operating or owning this aircraft is not covered by the warranty. All damage of this aircraft resulting from Acts of God may be covered by the warranty: the judiciary branch of the law or military justice system, a qualified aircraft mechanic or insurance agent specializing in this type of aircraft will ascertain the consequences resulting and/or damage from an incident and this company will enforce the invalidations in the warranty. All ensuing conduct and liability of the owner of this aircraft resulting from purchasing this type of aircraft is the responsibility of its owner. Contact information: ....

Activities and Incidents That Will Void this Warranty

For verification purposes for the invalidations determinations of the warranty for infractions of the law and infringements of legal precedence in the event of any qualified activity in that regard, interpretations will be made by a qualified law enforcement officer, aircraft mechanic, or insurance agent, who can specialize in this type of aircraft, especially for litigation purposes.

Any discovery of a warranty, certificate of title, or certificate of airmanship, or any other document made for any aircraft made by this company, found to be fraudulent, written for these types of aircraft, sold by this company, and/or found to be counterfeit shall not be valid, enforced, or enforceable, and shall be unusable to obtain any offers made by this company. The warranty shall be void for individuals subjected to and successfully prosecuted for fraudulent or counterfeit documents pertaining to the aircraft produced by this company for the duration of their sentence.

Any aircraft made by this company discovered in the United States, made to be fake, counterfeit, or fabricated by some other company with the intent to defraud against any of the purposes serving this warranty and to have services rendered to the aircraft in anticipation of having this warranty enforced, shall void this warranty. Counterfeiting, in the United States, of the type of aircraft made by this company shall endure for the duration of the patent period. Counterfeiting shall no longer be valid after the patent period ends.

A labor charge, compensation for maintenance, repairs, and replacement, will be all that is expected from this company in return for service and application of this warranty where this warranty shall be upheld and enforceable.

The following, will invalidate the warranty:

Any unauthorized modification, including modification with intent to defraud, made to the aircraft which does incidental damage to its original configuration. Authorized modification may be limited to approved store-bought items e.g. aftermarket pedals and seat, handlebars, crank arms, drive chain, chain-rings, hydraulic fluids, bearings, grease seals, lubricant grease, lubricating oils, and additions of approved accessories.

Using the aircraft for an illegal activity

Using the aircraft to do work which otherwise is intended to be done by some other device intended for the job, such as using the aircraft to trim a tree

Intentional hard landing: playing the game “Up the Chimney” will void the warranty. Up-the-Chimney is a vertical rapid-descent game where the player attempts to stop the rapid descent in time to just touch the ground without a hard landing.

Flying in to a restricted air space or “no-fly” zone will temporarily void the warranty. If an accident occurs while in the restricted air space or no-fly zone the warranty will be void.

The paint on the aircraft is not covered by the warranty. New original decals however, are covered by the warranty.

Theft of the aircraft is covered by the warranty through insurance coverage. Insurance should be obtained in the event of any likely theft of the aircraft. Damage or loss to/of the aircraft resulting from theft of the aircraft is covered by the warranty through the insurance company’s issuance of a damage claim after litigation is successful to obtain warranty coverage, and the insurance company’s use of the original warranty with respect to any aircraft produced by this company and the owner of the aircraft obtaining the warranty coverage to obtain warranty coverage for the aircraft shall be valid. This invalidation shall also include replacement, replacement parts, and vandalism of the aircraft and these stipulations shall not be excluded from this invalidation for warranty coverage of the aircraft. Any requested modifications to any aircraft made by this company being covered by insurance provided for by this warranty may be made at this time: however, the full purchase price for the modification additions and/or changes to the aircraft including the labor cost if it is requested by the owner to have the modifications assembled or disassembled to or from the aircraft will apply. Any permanent modifications to any aircraft constructed and designed by this company resulting in any positive and/or progressive improvement to any aircraft made by this company will result in a discount coupon being given to the owner of this aircraft to which the modification was originally made which will be valid indefinitely towards the purchase of any new aircraft produced by this company. Any and all modifications to this company's aircraft which are recognized and acknowledged by this company which shall improve the quality and performance of this company's aircraft and are implemented into production shall be protected by the warranty and shall be the property of this company. Anyone submitting a design change or modification for any of the aircraft produced by this company expecting to receive a discount coupon for a change or modification to any of the aircraft produced by this company must be an owner of an aircraft produced by this company and the change or modification shall apply to the aircraft owned presently; proof of ownership is required.

Any aircraft made by this company with respect to which a change or modification is requested found to be an aircraft taken into possession by theft by any means or by any means whatsoever whether by the individual or individuals requesting the change(s) or modifications or someone else or others, shall have their aircraft secured, and the discovery of the theft shall be reported to the proper authorities.

Fire, also, is covered by the warranty and is to include insurance coverage for fire damage and the aircraft and/or any of its components shall be replaced where an insurance claim is produced legitimately through litigation.

The electronics of the aircraft have their own warranties and are not covered by this warranty. Insurance should be obtained to protect the electronics, alternator, battery, and electrical wiring.

The pontoons are not made by Positive-Feedback and have their own warranty and are not covered by this warranty.

Carrying backpacking equipment and other items on or with the aircraft is at your own risk, or you should have insurance in the event that you could possibly lose any of these items while using this aircraft.

Failure to maintain, on a routine basis, the proper fluid levels in the reservoirs and to keep the chains oiled and the bearings greased otherwise resulting in a catastrophic failure of the aircraft. Always lube and oil the aircraft before going aloft if it has been left out in inclement weather.

Any legislation passed in to law which will force this aircraft to be forfeited to another owner by any means, or not allow the owner of the aircraft to be set aloft by its owner, or creates a constriction in the flight path of these aircraft or any circumstance by law which results in a more dangerous situation of operation of these aircraft by their owners, will void the warranty.

Any legislation passed prior to or after this type of craft is set aloft for its maiden flight, or before or after this warranty has been made available to the public, that harms the public in any way regarding their freedom of free range flight, safety, and welfare will void this warranty to any users who intend to operate one or more of this type of aircraft. Any and all warranties for this and all types of aircraft that are produced by this company that are owned by anyone who would do such legistation will be void on that account.

Failure to meet any requirements laid down by the FAA or legislation passed in to law by the Federal, State, and local Governments providing for freedom of free range flight, safety and welfare of owners of this type of aircraft operating their own personal aircraft of this type in any situation under any circumstances will void the warranty.

If a nation's Government, of belligerence, voids the warranty of the public by passing in to law any ruling harming freedom of free range flight, safety, or welfare of the public, the warranty of the public shall remain in full force and effect for the public of that nation and those nations Government's officials' warranties for these aircraft shall be void.

Owners of this type of aircraft may engage in combat with belligerent and hostile adversaries combatting against their home nation during wartime and this warranty will be in full force and effect for any occurrence resulting from warfare including the invalidations of this warranty. War is an Act of God.

Criminal friendly fire will not be tolerated and will void the warranty of the owner of the aircraft committing the criminal friendly fire, permanently and including all and any further purchases made of this and any other types of these aircraft produced by this company, for the remainder of the duration of the individual or individuals life or lives, forever. The criminal friendly fire does not have to come from a pilot or pilots of one or any of these aircraft; an owner of this type of aircraft or any of these types of aircraft produced by this company, committing the criminal friendly fire will suffice to satisfy the warranty invalidations. Any criminal prosecution for criminal friendly fire shall successfully include all parties involved and successful prosecution for the above invalidation claim will void the warranty of each individuals account.

Territory or possession of, for ownership of the atmosphere of the planet is outside the scope of this warranty. Territory is made of earth and objects which project out from it and it has boundaries which exist at the perimeter of the ground of which it is made and its limits extend only to the outward appearance of the protuberances which project out from it. The limits of privacy recognized by this warranty: everything that isn't the common air that is within bounds of one's own property perimeter constitutes ones privacy. Privacy extends to the tips of ones fingers as far as one can reach. The Troposphere is the habitable zone in which these types of aircraft may be found aloft, if not otherwise, and that includes the altitudes from sea level to their surface ceiling. Any presumption of ownership of the common air or possession of territory within its realm is excluded from consideration by this company and by this warranty, and any consideration, whether by individual or entity of entitlement to forge operations in this country, for ownership or possession of the common air brought up in litigation by a or any plaintiff, possessing an aircraft of this type produced by this company or any aircraft type, shall void the plaintiffs warranty.

No other warranties but this warranty, for the types of aircraft made by this company shall be valid, enforced, or enforceable and any attempt to use such warranty shall be null.

Human powered flight is at your own risk. Pay attention and fly responsibly.

Crimes instated after-the-fact shall void the warranty as they manifest themselves in the laws over time and this warranty shall remain in full force and effect. Any law or laws made or removed that affect these types of aircraft and their operation shall affect this warranty and the warranty shall reflect such changes.

Once the aircraft produced by this company is being sold for parts or has been scrapped this warranty is no longer valid for the individual aircraft serviced by this warranty but the warranty will continue on the flying aircraft produced by this company, with the used parts.

-These invalidations are acknowledged Worldwide by this company.-

This warranty is subject to change without notice.

Human Powered Transportation Means

Human Powered Transportation Means

As the pages following progress, initially the calculations show the diameter for the contact radius from road to the axis of the axle. The applied force at the road/tire contact area is surmised at 1500 pounds and the static calculation derives 9750 pounds at the impeller blades for the differential impeller. A top speed is surmised at 230 miles per hour and the conversion factor to feet per second will allow the circumference of the tire to derive the revolutions per second, or two crank revolutions per second at a cadence of 128 per minute. The hydraulic fluid density is 53.9 lbs/cubic foot. Calculating the volume of the impeller gallery gives 1.40359 cubic inches. Dividing by 12 cubed cubic inches per cubic foot leaves a dimensionless number in the equation for the which the exponent is found, the fold for the hydraulic fluid which is multiplied to the impeller gallery volume, the revolutions output of the rear wheels at 230 miles per hour, and by 2 because the impeller blade vanes volume displace half the swept volume of their full circuit of revolution in the impeller gallery.

Several corrections were made at the beginning of the design process. Eventually the secondary and primary power transmission gear depths were settled on and the radii for the primary impeller, and the resulting force at the operator's crank pedals was only 39.9291 pounds. It had initially been intended to be 40 pounds. Slip was never corrected for, so the gear depths in the drawings is shorter than it should be but not by much. The desired applied force at the crank times the length of the crank arm divided by the radius of a 52 tooth sprocket equals the resulting force at the chain-ring that drives the primary power transmission center gear. The hydraulic force at the primary power transmission mesh gear teeth times the quantity of the inside radius of the center gear of the power transmission times the applied force at the radius of the driven sprocket driving the power transmission center gear, close quantity, equals the radius of the driven sprocket of the power transmission center gear driven by the crank sprocket. The vehicle may go a little slower than was actually intended without the slip being calculated in but not much slower.

I began to draw tubing, and sprockets,...and more tubing. I also managed to correct my equation for calculating the shear force in tubing and will have included it on the next page.

While finishing the assembly file the frame drawing somehow obtained open loops and intersections in the primitives and I was not able to produce an extrusion and could not find the open loops or intersections.

I made lists of items to draw. I made a list of the last things I had to draw. And the last thing I started doing is an attempt to design a backpacker's self-propelled power unit using my self-propelled motor, attempting to make it as small as I can make it: however, there are no included drawings or arithmetic for a final product. Then I started designing a truck cab for an 18 wheeled tractor trailer. I made a zero error in the ones place and the mathematics returned 10 times the required product. I left the error and considered the tractor pull contest at the Fairgrounds. The tractor trailer cab is mathematically originally designed to maintain 75 miles per hour at 6% grade with 30 tons towing.

Shear area formula:

$$(\text{pi} \times r_2^2 - \text{pi} \times r_1^2) \times \text{shear force of desired material in lbs/}^2\text{in.} = \text{required force in pounds.}$$

$$r_2 = \sqrt{\frac{\text{required force in pounds}}{(\text{shear force of desired material in lbs/ in.} \times \text{pi})}} + r_1$$

$$r_2 \times 2 = \text{diameter.}$$

Mathematics used in the formulation of equations (Science of Mechanics):
Basic Functions
Algebra
Geometry
Trigonometry
Science of Mechanics (Fluid Mechanics of Inertis)

The Science of Mechanics appears similar to calculus but is only algebra. The Science of Mechanics implements resolving the exponent of any base with long-hand mathematics, and the resolution may be observed unlike observing the resolution for solving an exponent using a scientific calculator.

"The Formula for Fluid Mechanics of Inertia" is also a basis for analytical brain function, the likes for which how chemicals can make value changes to variables in soma clusters and develop a tangible solution to the brain creating information that an individual understands is still beyond the scope of understanding. A soma contains a variable, a soma cluster is an equation e.g. smell, or taste etc. Information enters the brain through its senses sending chemical and electrical signals to the soma clusters and the brain is continually making variable adjustments to the values in the somas and the solutions are continuous and uninterrupted. The frontal cortex also conatins soma clusters which operate in the same manner using processing of information gathered throughout ones life. The chemistry of the frontal cortex however, is beyond the scope of this text.

The Following Pages Contain The Specific Initial Drawings and Calculations for the Human Powered Helicopter Final Draft

Calculate where the center of the lift blade and tail rotor blade vanes are, balance the weight on the AA'. Offset the blade vane mounting and center to axis

$$\frac{262.65723 \, in^3 \times 165}{12^3} = 25.0801$$

$$\frac{.53383 \times 165}{12^3} = .05097 \, lb$$

11.84 lb

25.13209 lb

$$\frac{51 \times 165}{12^3} =$$

.74995 al

4.86979 al

.04961 lb air

97.85398

42.92699

$$\frac{54.92659 \, in^3 \times 165 \, lb/ft^3}{12^3 \, in^3/ft^3} = .03179 \, lb.$$

.00216 lb air

.00384 lb air

895-1213

6.6 lbs

.04961 lb.

.00487 lb.

328.35 in³

256.19291

.93 in forward

9,668

438

329.5 3in × 165 lb/3in / 12³ 3in/3in

67.368 × 2

× π = 423.286 circ

31.463 lbs × 2

Assuming the lift blade reaches its weight in impact reaction force in

360/48 = 7.5°, and the maximum lift blade revolution rate is 9.5017 Rev/s

$2^{5.5}$ × 48 × 9.5017 Rev × 31.463 lbs = 14,349.7 lbs of centrifugal force at the lift blade hub bolts / 3 bolts = 4,783.232 lbs each bolt.

A .1875 r = .375 diameter bolt exceeds the needs .1875² π × 60,000 / 2 in

= 6,626.8 Pounds shear area of bolt.

tail rotor blade:

29.646 r × 2 × π = 186.27131 in / 25 = 7.45085 in,   711.2098 lbs   11.2549 Rev/s

48 × 11.2549 × 7.4092 = 4002.699 lbs centrifugal

77.5943 in × 165 lb/

(1646.2553 lbs / (7.67613 in × 69.5 lb/ft − 2 in / 2 in) / 12³ in/ft) + n 5.4917 in + 2.1844 in = 7.6761 in

= 12.3018 × 7.6761 × 14.0569 = 1,327.3946 in

1327.3946 / 2.5464 in / 2 PTU / 8 mgT / 8 Rev. = 5.43 in. gears depth at secondary tail rotor PTUs @ 8 Rev. per 2 crank S/R.

11.93[illegible] in + 4.7615 in = 16.6959 in

(171.0762 lb / 16.6959 · 69.5 / 12³ − [illegible] / 2 in) + 11 = 7.9903 · 8 · 16.6959 in

= 1,067.2482 in / 2.5464 in / 4 PTU / 12 mgT / 2 Rev/crank S/R = 4.3658 in gears depth tail rotor primary PTU

What is the fastest cadence that can be cranked with a [illegible] load acting on the pedals?

~~2098 lbs~~

$$\frac{3689.53761\ \text{lbs} \times .93\ \text{in}}{12.25} = \frac{280.10367\ \text{lbs} \times .93}{4.138}$$

$$= \frac{62.95225 \times 4.138\ \text{in}}{7\ \text{in}} = 37.21377$$ tail rotor applied load at the crank pedals.

$$\frac{8188.5559 \times .93}{25.5} = \frac{298.64145\ \text{lb} \times .93\ \text{in}}{4.138\ \text{in}} = \frac{67.11855 \times 4.138}{7}$$

$$= 39.67665$$

tail rotor blade + lift blade

$37.21377\ \text{lb} + 39.67665\ \text{lb} = 76.89042\ \text{lb}$ at the crank.

1 bladevane volume (tail rotor) = 24.7086 3in 90 3in × 2 = 180 3in

123.5432 3in

1,169.4584 3in swept vol. T/R 8517.06207 3in

$\frac{6188.5559 \text{ lb} \times 15.775 \text{ in}}{240 \text{ in}}$ = 538.227 lb + 172.9828 lbs. =

711.2098 lbs.

$\left(\frac{\frac{538.227 \text{ lb}}{123.5432 \div .0612 \text{ lb/3in}}}{12^3 \text{ 3in/3ft}} - \frac{2^n}{2^n}\right) + n$ = 16.877

$\left(\frac{1169.4584}{123.5432} - \frac{2^n}{2^n}\right)$

= 3.1832

16.877 − 3.1832 = 13.6938 revs. per 2 crank C/R

123.5432 ·

$\left(\frac{\frac{711.2098 \text{ lbs}}{123.5432 \cdot .0612}}{180 \; 123} - \frac{2^n}{2^n}\right) + n$

~~16.877 − 3~~

16,70231 5.44741 ~~16.70231~~

17.2801 − 3.1832 = 14.0569 revs per 2 crank C/R per blade ← 11.2549

$\frac{711.2098 \times 13.165 \times 29.505}{5.6875}$ = 1,646.2553 lbs.

3,689.53761 lbs

blade vanes volume = 1934.3371 ^{3}in

.07651 lb/^{3}ft = air density @ ~~sea level~~
.0612 1000 ft 5??

$2.394\text{in} \times \left(\left(\frac{189.55}{2}\right)^2 - \left(\frac{31.55}{2}\right)^2\right)\pi$ = swept volume = 46,938.3913 ^{3}in

1255
950
500 lb

$\left(\frac{1934.3371\,{}^3\text{in} \cdot .07651\,\text{lb}/{}^3\text{ft}}{\left(12^3\,\text{in}/{}^3\text{ft}\right)} - 2^n/2n\right) + n = 12.1253\left(\frac{46,938.3913}{1934.3371} - 2^n/2n\right) + n$

= 13.7816
13.6925 = 4.5166 = 2^u
- 5.8958

15.3975 = 7.8858 Rev
7.7967 2 crank C/R

$\frac{1934.3371}{5} \times 2 = 773.7??$ ^{3}in

$\frac{660.696\,\text{lb} \times 67.368\,\text{in}}{10.325\,\text{in}} = 4,310.8734\,\text{lb}$

= 5.8958 = 2^u

$\frac{1110.696\,\text{lb} \times 67.368}{10.325} = 7,247.009\,\text{lbs.}$

$\frac{1467.6668 \times 67.368}{10.325} = 9,576.152\,\text{lb}$

~~555.348~~

1110.699
+
1467.6668

15.3975 - 5.8958 = 9.5017 Rev.

$\frac{1255 \times 67.368}{10.325} = 8,188.5559\,\text{lb.}$

$$\left(\frac{\dfrac{9576.1527\,lb}{69.5\,lb/_{3ft} \cdot 7.70493\,_{3in}}}{12^{3in}/_{3ft}} - \frac{2n}{2n}\right) + n = 14.8861 = P'$$

14.6128

$$\left(\frac{\dfrac{9576.1527\,lb}{69.5\,lb/_{3ft} \cdot 6.1556\,_{3in}}}{12^{3in}/_{3ft}} - \frac{2n}{2n}\right) + n = 15.1804 = P''$$

15.0094

$7.70493_{3in} \cdot 14.8861 \times 7.8858_{Rev} + 6.1556 \cdot 15.1804 \cdot 7.8858 = 1641.3634$ (above the line: 6128, 9.5017, 0094, 9.5017; at right: 1947.6777)

$3in$

Calculate the Primary impeller meshing gearteeth (mgt) Power transmission unit (PTU) gears depth for 15 teeth-center gear: To output 1.C4 impeller gallery fluid volume.

$$\frac{1641.3634_{3in}}{\dfrac{7.6391^{2}_{in} \times 8_{Rev}}{2/3}} = 13.429_{in} = 5.3117_{in}$$

Primary impeller at 8 Rev's.

$\pi r^2 = \frac{1^2 in}{~~60,000 lbs.~~}$ $\sqrt{\frac{1}{\pi}}$ $\sqrt{\frac{60,000}{\pi}}$ 1.1284 in Ø ~~141~~

412.8329
487.6177 lbs.

$\frac{18.5228 \text{ in}^3 \cdot 69.5}{12^3}$ $-2^{11}/2n)+n \times 8 \times 18.5228 + \left(\frac{487.6177}{\frac{9.8666 \cdot 69.5}{12^3}}\right.$

9.2784 9.0823

$-2^{n}/2n) + n \times 8 \times 9.8666$ $= 2{,}180.0101 \text{ in}^3$

10.2

2,136.42

~~61965.153~~

$\frac{\frac{\frac{\frac{2180.0101 \text{ in}^3}{2.5464 \text{ in}^2}}{8 \text{ PTUs}}}{2 \text{ Rev}/2 \text{ crank cir}}}{12 \text{ mgt}}$ (meshing gear teeth) $= 4.4589$ in gears depth

4.3698

311

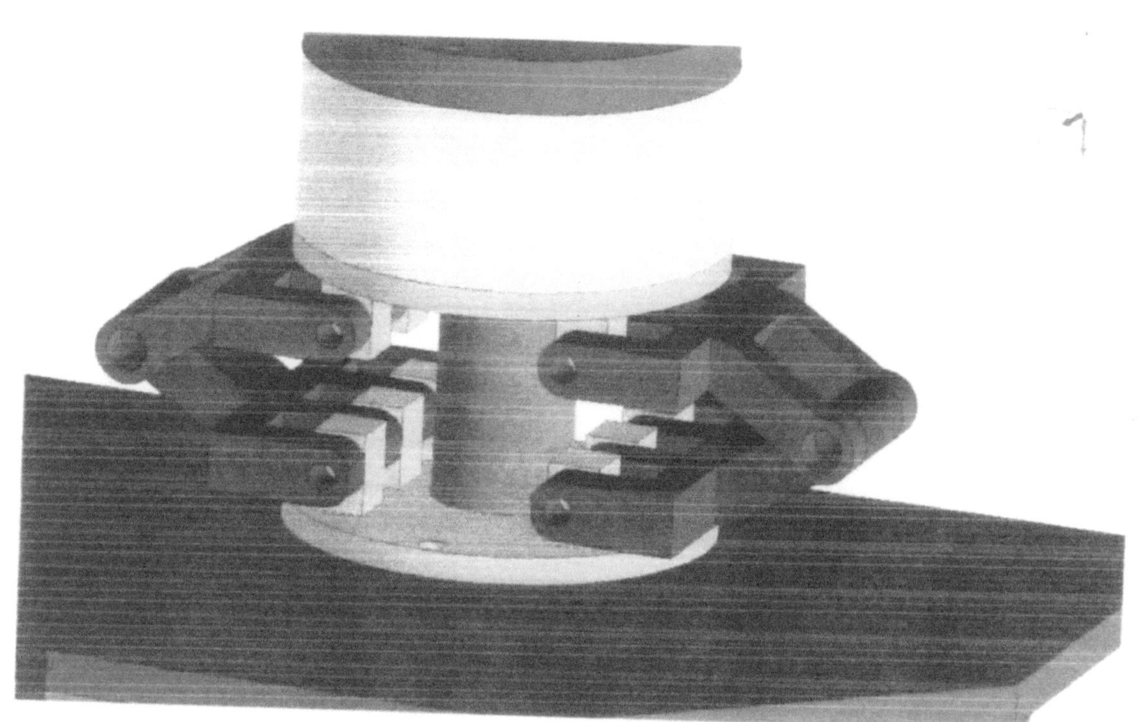

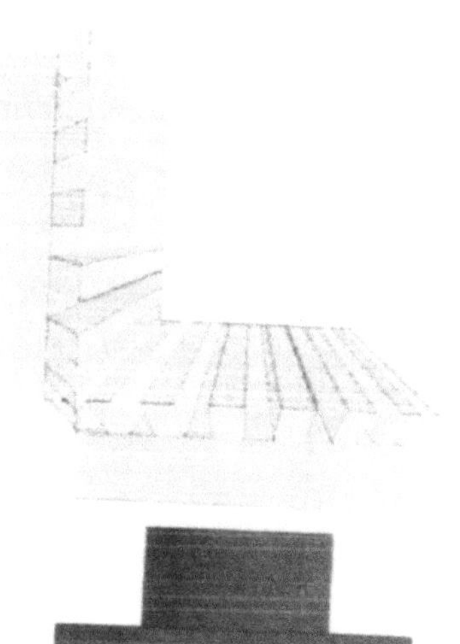

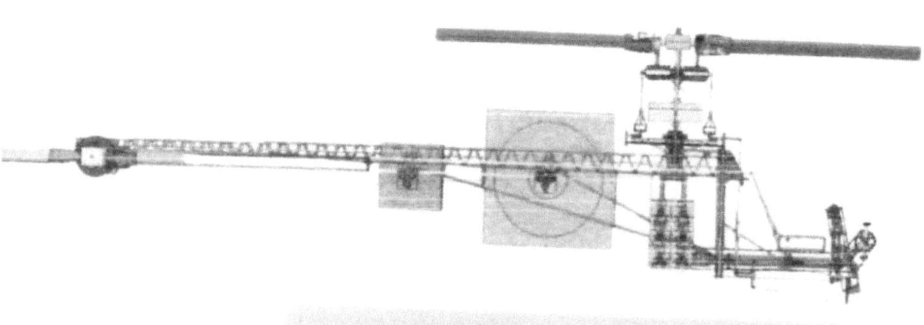

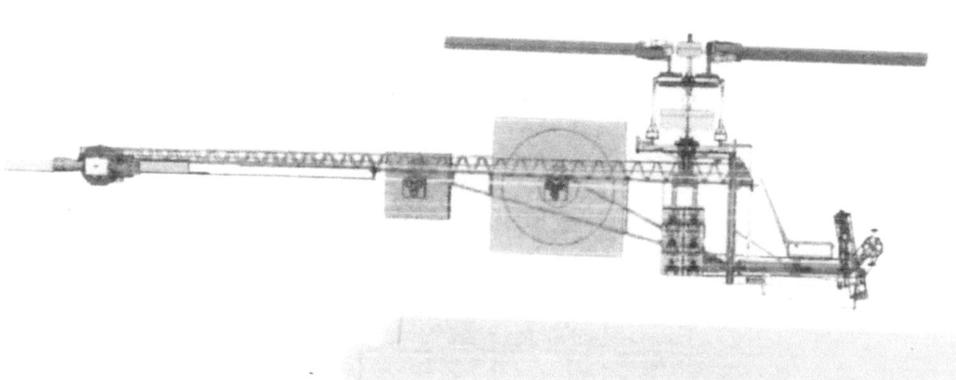

108,698

Remaining chain drive 20.711

3,66275
,101057

,148294
68.493247° from
,128712

tubing

.725 ↑

.599

13.598

holes

3.887

tracks

3,429101
,185526

1,087888

1.767

31,210922

24,316852

2.356 longer

1.75 75°

125

,042
,206
,10307

14,029394

2.113

3,7606

3,184

(.043)

1.546

~~3,1285~~

3.932

10,279

How many degrees does the flywheel rotate at 1 rpm and its impact reaction force is equal to its weight?
What is the cadence? 128 or sprockets ratio inverse/min.
Multiply by the sprockets ratio.
What is the density of the flywheel and what is its volume?
What is the center radius of the median circumference of apply reaction force?
[illegible] its applied force at 128 cadence equal to the [illegible] force [illegible] at the crank pedals.

$\frac{52}{16} \times 6°$ $\frac{54}{16}$

assuming:
If 19.5° at 1 rpm ^rotation of the flywheel develops the E/O r.f. (equal and opposite reaction force) equal to the weight of the flywheel then 1/120 of a minute will result in the E/O r.f. = 120 × the weight of the flywheel. The applied force at the pedal is static elastic equilibrium throughout the torques from the pedal center at the crank arm radius to the flywheel radius center at applied force =

at 1 rpm $\frac{52}{16}$

[illegible] force × crank arm radius / sprocket radius chainring × flywheel sprocket radius / flywheel center of applied force =

6° at the crank [illegible]
develops [illegible]

or use a known density and calculate the flywheel radius.

77 lb x 7" inches / 4.1375" r sprocket inches × 1.29324 inches flywheel sprocket radius / flywheel radius at center of applied force inches = product force at attempted radius.

make inches attempt.

product force [at the flywheel radius attempted ...]

lb. ft

165 lb/3ft

Material density lb/3ft × calculated volume 3IN (estimate .25 3ft) at the attempted radius / 12^3 3IN/3ft

Product force [at the flywheel radius attempted] × attempted radius

= required radius with known material density.

The volume of the calculated volume remains the same volume at the required radius.

$\frac{165 lb/ft^3}{13.82228 lb} \times .25 ft^3 \times \frac{12 in}{1 ft}$ = ~~35.81174~~ ft.? 2.98431 ft r resulting radius.

increase

$\frac{\text{Resulting radius (in.)}}{\text{Desired radius (in.)}}$ × attempted cubic volume = required cubic volume ...

$\frac{2.98431 ft \times 12 in/ft}{7 in \text{ desired radius}} \times .25 ft^3 = 1.27899 ft^3$ in. ... radius.

.00006 ft

× 12^3 in/ft

2,210.9472 in

area × 4D = volume attempt. ft or in, whichever.

$\frac{\text{attempted volume}}{\text{required volume}}$ × attempted ~~radius~~ area = required ~~radius~~ area.

1 in²

1 in × 14 in × π = 43.9823 in attempted volume

$\frac{43.9823 in}{1.27899 \times 12^3} \times 1 in^2 = .0199 in$

$\sqrt{.0199}$ = .14107 in on a side square.

.14107 × 14 × π = 6.20447 in

6.20447 in × 165 lb/ft ... = 592.44 lb

.159244 lb × ~~1.20 impact reaction~~ sqft

$\frac{360}{19.5} \times 2 \frac{\text{crank rev}}{\text{sec}}$ = 36.92308 swept angles of [illegible]

.159244 × 36.92308 = 21.87471 lbs.

$\frac{21.87471 \text{ lbs}}{\frac{165 \text{ lb}/ft^3}{12^3 \text{ in}^3/ft^3}}$ = in^3

.19107² × 14 in × π × 165
= .87528 in^3

$\frac{.1876}{12^3 \text{ in}^3/ft^3}$ × 165 lb/ft³ = .08358 lb.

$\frac{2210.09472 \text{ in}^3}{12^3 \text{ in}^3/ft^3}$ × 165 lb/ft³ =

77 × 7

$\frac{360}{19.5}$ = 18.46154

operator force lb × 7 in × $\frac{1.273 \text{ in}}{4.138 \text{ in}}$ = flywheel [illegible] force inches.

(volume × density) (of known material) in^3 in

$\frac{144.71223 \text{ in}}{7''}$ × → at length 7.12 cubic volume

$\frac{21.87471}{165}$ × 12³ = 229.08787 in^3 → variable → $\frac{7 \text{ in} \times 165 \text{ lb}/ft^3}{12^3 \text{ in}^3/ft^3}$ (lb.) = 7.5806 ~~inch~~

$\frac{229.06787 \, in^3}{7.58026 \, in} = 30.22162 \, in^2$

$\sqrt{30.22162 \, in^2} = 5.49742$ inches [illegible]

$\frac{30.22162 \, in^2 \times 7.58026 \, in \times 2 \times \pi}{12^3} =$

3.91″ 9.41″ 137.5 lb. flywheel dead weight

$\frac{229.06787 \times 165}{12^3}$

$\frac{52}{16} \times 360 / 19.5 = 60$ $\frac{137.5 \, lbs}{60} = 2.29167$ lbs. (flywheel weight) at 120 cadence.

$137.5 \times \frac{7.58026}{1.273} \times \frac{4.138}{7''} = 484.$ [illegible] force at the crank.

$\sqrt{\frac{165 \times 7.58026 \times 2 \times \pi}{77 \, lb.}} = 10.10219$

$\frac{77 \times 7 \times 1.273}{7.58026 \times 4.138} = 21.87172$ lb flywheel dead weight

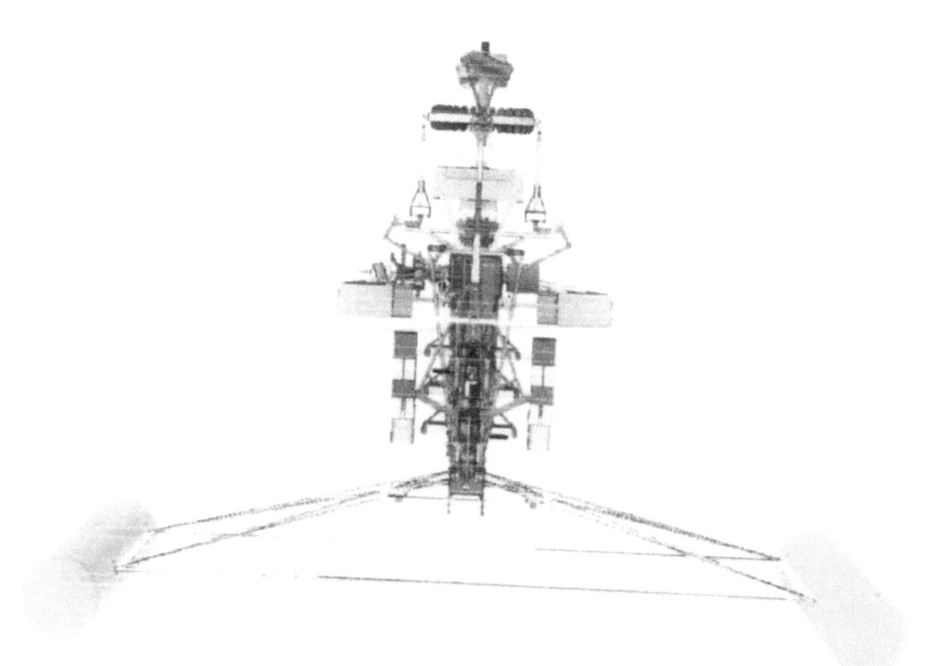

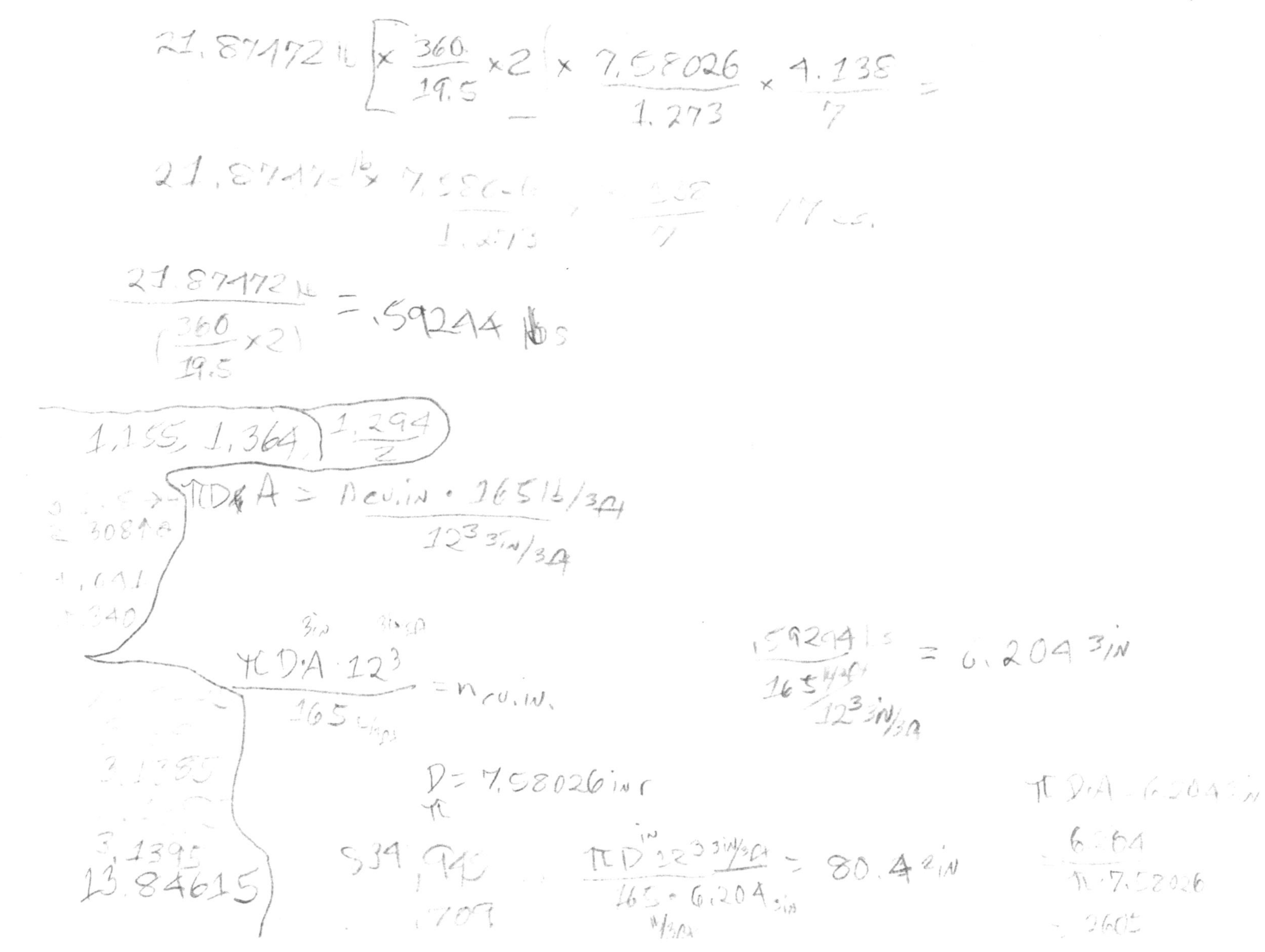
= .59244 lbs
1,155, 1,364
D = 7.58026 in
= 6.204 in
13.84615
= 80.4 in

11.921186°　　1.376
2.5　　16x6

tubing, primary
grease fittings
~~Seat belt mount~~
all weather cover
Cockpit control linkage covering
~~ferrying chain~~
~~secondary pivot end plates~~
frame to lift impeller gallery lacque brakes
~~Pilot bearing covers~~
Secondary tubing

$\pi DA = V$, $\sqrt{\text{area}} = \text{SIDE}$

Known density = 165 lb/3ft

Desired radius = 7 in

6° @ crank = $6 \times \frac{52}{16}$ = 19.5° at flywheel

77 lbs @ crank operator's force (varia...

2 crank revolutions/second

flywheel ~~force~~ weight × 19.5 × 2 = 77 lbs @ crank arm radius

flywheel weight = $\frac{77}{(19.5 \times 2)}$ = 1.97436 lbs. @ 7 inch radius

~~1.97436 ×~~

1.97436 × 19.5 × 2 = 77 ...

static force.

$\frac{1.97436 \text{ lbs}}{165 \text{ lb/3ft}}$ ~~× 12~~ = .00001 3in

12³ 3in/3ft

14 × π × 1² = 43.982...

360/11.

18.46154

Static force × $\frac{52}{16}$ × 2 × 19.5 − 77 lb

1.97436

= 173.25013 lbs.

$\frac{165 \text{ lb/3ft}}{12^3 \text{ 3in/3ft}}$ × n 3in = 1.97436 lbs.

n = 20.67693 3in

1.97436 × $\frac{52}{16}$ × 2 × ...

= 173.25013 lb

$\frac{20.67693}{43.9823}$ × 1² = $\sqrt{\frac{43.9823}{20.67693}}$ = 1.45846'

$\sqrt{\frac{20.67693}{43.9823}} \times 1^2$ = .68565 in on a side

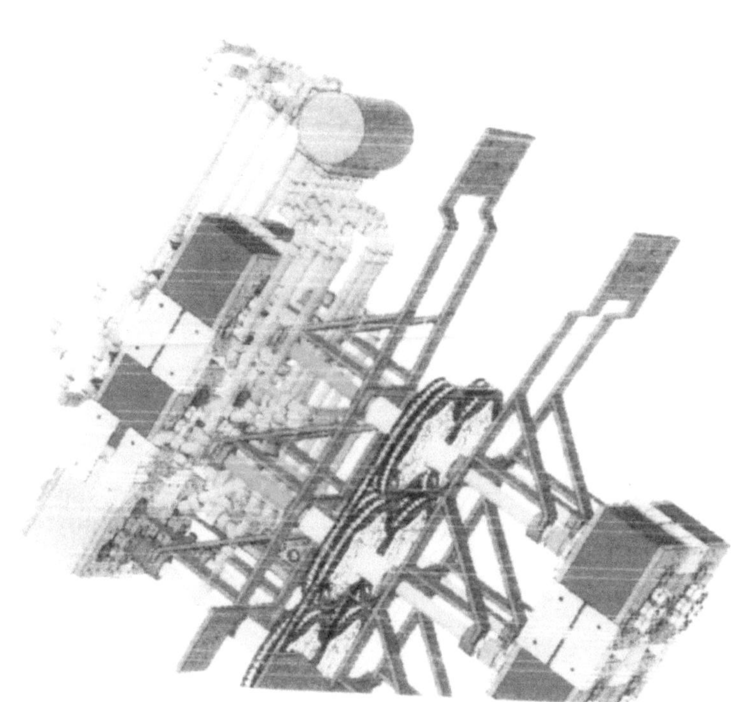

11.0513 + .25 = ... (11.36718 11.3638 · π · .375

− 11.05468 11.0513 ·π · .375) · 8 rev/2 crank C/R = 66.03783 3/in × 3

= 198.11349 3/in / .68662 in = 3.006" max depth before slip primary PTUs

4 PTU / 12 mgt

2 rev/2 crank C/R

1500 / 750 / 53.9 × 8.25224 −2"/2"+11 / 12³

=

2913.69705 −2"/2" +11

standing load 11.4227 ×

8.25224 × 8 rev. vol. ×2 the whole shebang

12.4227 × 8 × 8.25224

primary PTUs 1640.24163 820.12105 3/in 754.10288 3/in / .68663 in × 538 crank C/R

16 PTU / 24 mgt = ... × 9.111 = 56.68 ... gears

1500 lbs × 1.70072 / 7 in = 9.111 ×

28.35422 in. depth

3.006 × 9.111 = 27.388 in gears depth primary PTUs

PTUs 4×4 chain rings left out 10 lbs

9750 × 1.70072 / 11.05468 × 1.70072 / 18.85968

× 2.06901 ← 52t sprocket radius / 7 crank arm radius = 39.98106 lbs. at the operator crank.

474T · .25 / π / 2 = 18.85986 radius sprocket at primary PTUs (4×)

52 × .25 / π = 4.13803

4.13803 · π / .5 = .6T

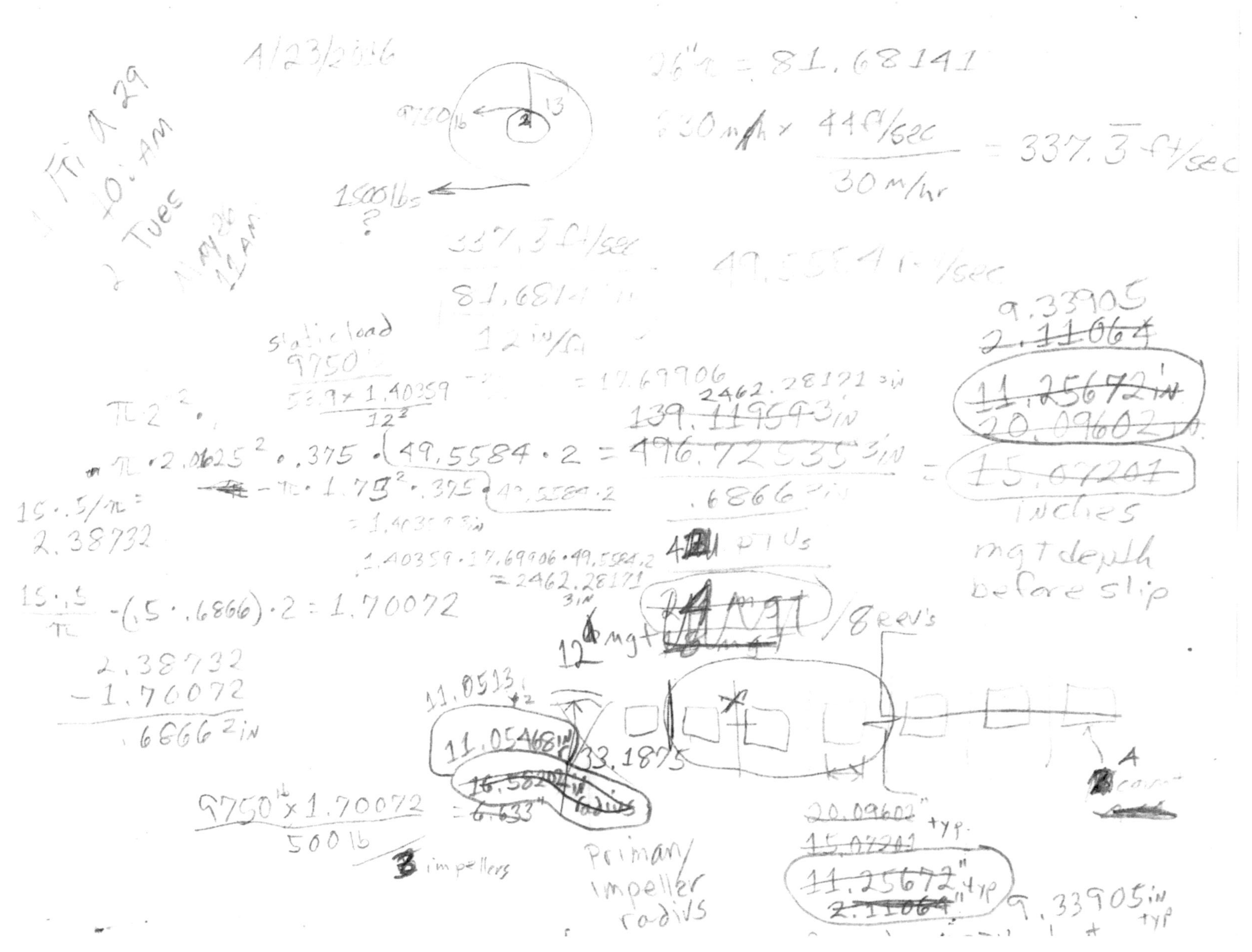

Correct for slip

$2^\circ =$ ~~13.88448~~ "15.658"

$3^\circ = 15.119$"

$1^\circ = 16.215$"

$1^\circ = 14.296$

1.385° no change

1.485° wrong way

$1.285^\circ = 16.026$

$\frac{16.026}{-15.998}{.028}$

$1 + \frac{.002}{.028} \times .1 = .07143$

$1\ \frac{1}{557} \times \frac{.215}{.557} = .386$

$1.386^\circ = 15.998$"

$1.386 + .07143 =$

$1^\circ = 16.215$

$2^\circ = 15.658$

$\frac{.215}{+.342}{.557}$

$\left(1 + \frac{15.658}{16.215}\right) \times 1^\circ = 1.96565$

$1^\circ \rightarrow 2^\circ\ \ 1^\circ$

$= 1^\circ$

$\left(1 + \frac{.002}{.028}\right) \times 1.386$

$= 1.485^\circ$

$.557 \times .215$

still yes!

$\frac{1.285}{1.07143} = 1.38386^\circ$

$\frac{1.285}{1.07143} = 1.2936$

$=$

Springs
Shockabsorbers
Powerplant
Wings
~~Throttle~~
~~Wheels & Tires~~
~~Brakes~~
~~Seat~~
May need a flywheel at the crank
~~Straight rack~~

12T @ .25
1.03451 D .69121
Rear axle to 2" dia.
Only U-Joint + steel,
Rear end Aluminum 6061 T6

upside-down

$\pi r^2 \cdot 60{,}000$ lb/in = 10,000 lbs

$\sqrt{\frac{10{,}000 \text{ lbs}}{60{,}000 \ldots}}$ ×2

= .46066" diameter

$\frac{10{,}000 \times 1.70072}{.375}$

= 45,352.53

.98103 in diameter

Primary PTOs use aluminum shaft to main drive sprockets

.160m

.375 dia.

$\frac{12 \times \frac{.425}{1.5}}{\pi}$ = .75

$\frac{.75\pi}{1.5}$

Martina

1/4 lb smoked Bacon cook down
1lb pork sausage browned
chopped return bacon to sausage
take sausage and bacon out of pan
4 tbsps sweet butter
[illegible] small onion
soften onions
2 tbsps minced garlic

roux
4 Tbsp flour little at a time, mix in to onions + butter
slightly brown flour

2 cups half & half work in
chicken bouillon cube ground up work it into mixture

1 tsp sage
1 tsp thyme
1/2 tsp cayenne pepper

Salt & pepper to taste
1 tblspn Worcestershire
fresh chopped parsley
Green onion

add more 1/2 & 1/2 [illegible]
[illegible]

Chain drive
Primary tubing, head, draw, draw collectors
Secondary tubing, head reservoir
Secondary PTU moving parts, external parts
Secondary PTU Secondary tubing, head grease seals, collectors
grease seals, primary PTU, primary
Secondary reservoir
Secondary collectors, PTU assy.
head, final tubing draw, collectors
primary PTU gear side

11.716
9.142
9.338 Wings
11.392 Flywheel?
26T sprocket .5
26T sprocket
19T sprocket

Ext .164 .114
.200 Pitch WAS
74027
5590 1000 63501
T20
12462734°
310875

30.664
− .5

seq name
Pedals
Puddles
152361259
3250
33
1.33r
.855
1.066
1.593
2.066
1.586

10.0"
3.25
5.636
5.537
14.86 11.527
2.541

4.354 4
4.854 7.333
5.358
0.500

9.000"
8.25
— 2.00
.6165
43.251035°

583.5035 3.008
.5525
—1.009
2.209293°

33.66
20.573117°
32.464934°

.5825 6.1875
1.1875
3.001 2.5 1.759 = 2.824

1.231
— 2.075 .5825
4.075.
5.25
2.01

1.375 : 1.316 :: 1.3125 :
1.1675
— 4.675
9.00 2.25

1.375 1.316 :: x : 1.1345
1.1355
— 1.066 1.172

1.231
2.025
2.396
11.25

12.754367° 3.00

3.073
3.573
4.074

21.716
12.00 .5515
6.00 — 3.009
5.375 2.26
.582
— 1.009

4.851
4.858
5.358
1.086
2.086

19.34846°
.5525
3.008
— 1.009

1.133
1.007
.5515
.048 3.009
1.007
1.079
9.798
65.755043°

1½ ↑ 23.639 / 2 ↓

−1" ↓

−3.009
2.5265
4.981 up
4.231 33.959

← .5525
− 3.0081 − ½
8.906
8.156

4.3065
−4.075
6.731
5.981
61.484401°
28.515599°

8.10189°

−3.008
7.9225
3.606
38.646

35.499824°
39.500176°

2.434 inward
13.317 / 2

.6165
− 2.00
8.906 *
8.156
18.1 4032°

1.74
− 1.066
8.906
8.156
1.1675

17.292 / 2
− 4.075
12.754364°
1.318

5 ← into center

37 up
× 5" tall

.5515
− 3.009
11.906
27.2925 3°

6.1055
− 2.000
5.702
17.093197°
10.718242°
30°
31.465742°
− 2.075

21.048882
1.244942 √34°
.744942 → 1.0855
1.51066
4.702213°
12.281
11.531

.6145
− 2.00
11.156
32.51598°
4.239966
24942°

14.177

13T
14T

.925039
1.124329

21.876 + .05 + 1.1

5.77873
33.387766°

1.231
1.028899

.75 SD. 1.000 OD W
SCE - 1211 .0874
.787 ID
1.85 OD
.551 depth

16.125 − .125 − .6866 = Dia.

6.625
3.375

13.563

21.688

4.075
6.1415
6.1731
5.232
~~61.154441°~~
63.710014°

26
3.1085
18.324
7.026
120.094064°
45.907548°
45.872477°
38.552716°

28.176571°
68.292491

3.1415 →
20.399
8.776
8.026 → 15°
100.460198°
1.239802°

4.9225
19.332
16.526
9.776
76.669798°
79.529896°
46.607782°
55.506067°

head inside bottom 3
draw inside bottom 3

14.0 9.798
42.356
21.178 = 2

7.3345
18.324
7.026
4.00 → 2
55.991224°
8.15994

7.3065 →
20.399
8.776
3.5 → 4.0
6.1085
18.324
10.526
7.00
63.368783°

86942
72.329779°
73.209212
75.262047°
79.402637°
53.089787
55.952438°
65.049742°

~~Adjust main frame tubing~~

rack and pinion casing + mount to frame

gusset plates in main frame at intersections

22.481928°

3/4" chain links

.375 chainring depth

13.202

18.51

.685603

1.369139

.542

101.317421°

1.6727

76.

10.02809'

1.637725'

21.01841°

1.5085

Underside frame to secondary PTO

.3125

.8039

.7969

83.309521°

.625

.841↑ + .391

.206← + .095

4.24→ + 1.0 *

51.11999

77.02234°

26.4445

12.4765

10.2285

7.654→

1.232↑

7.34→ 8.872→

.222↑ 2.298↑ 8.414

.222↑

.885→

— 2.762

8.256→

1.268↓

1.486↓

2.387→

2.488

20.596388°

2.576639°

1.89955

6.97355

.0456

.0205

.0645

.0175

.0055

.916←

←.23

↓.13

fastenal.com

12.189↑

7.051←

1.268↓

.301→

10.227038°

3.937608°

.224742

.com/pay

x-prize vehicle 52.48624°

304
305
321
333
335
338
350

337 pc

25.6945

2675713

Titanium

550

Human Powered

Shark Motorcycle Audio

22050

.54127
.5469
1,0595
1,0665
1.3125

8,256 → 1.5625
1.268 ↓ 1.8281 +.007
1.8351
2.0781

HP60

~~Education~~
~~Experience~~

.541 ↗
1.194 ↘
~~Anchor~~

Date
Copyright

1.348
1.6094

3.125
4.601 ↑

.773

Clothes Donation 1.5625
2.0781
1.828
1.8351

Ceramic shell casting

37.677
4.4531

$\tan^{-1} \frac{11.628}{42.842}$

YYA

Bags

.084 ↓
.399 ←
+2.8281
~~25.1444~~
22.4990

25.989661°

3.3515

3.382

2.625

(15 .3125 $^2\pi$ − $5^2\pi$) · .375 × 53.9 lb/ft³ × 1 ft³/12³ in × 1/2

.3906 .23685 lb/2 = .11842 lb

(x) 16 1.8906 = 1.89479 lb/2 = .94739 lb × 5"/.13926r = 4.25191 lb

2.8281$^2\pi$ − 2.5$^2\pi$

.906 .05178 lb/2 = .02589 lb .94739 lb × 5 in r / .13926r = 34+ lbs.

60.74564 3in

5.77719 3in .08582 2in 60.74564 3in / .34331 2in =

.25749 2in

1.88518 3in .34331 2in

.13261 Dia. = 11.05882 in

.03444 2in .10709 lb of hydraulic fluid in Mgt

.11139 2in 1.85911 lb force at Mgt

.14583 2in

3.951

1.

$(1.9755^2\pi - 1.5^2\pi) \times .538 = 2.7931[illegible] \times \frac{4x}{8x}$

$= 22.3454[illegible]$ 3in

$\frac{53.9}{12^3}$ lb/3in [illegible]

$\frac{22.34549 \text{ 3in}}{.583312 \text{ in}} = 38.3081[illegible]$ in

$\frac{19.15405}{2 \text{ ptv}}$

$= 9.57703$ in each [illegible]

$= \frac{9.57[illegible]}{2 \text{ pt}[illegible]}$

$= 4.78851$ each pt[illegible]

5.875
4.75
3.625
2.5
3.375
2.25
1.125
1.25

1.292
2.1625

1.00 OD
.03471
7.01 10.00
99 installation for phone
41 33 internal
20"

Secondary PTU chain cover plate
reservoir caps - threaded
Secondary PTU support underside
" " " lateral
all PTU spacers and bolt holes
Rand bar primary PTU underside support
frame bearing cover plate(s) and bolt holes
Resize aft air dam
Body panel cut-outs for lateral and underside supports
Adjust underside support to fit between gusset plates
* frame diagonals at front
Calculate diagonal at front top of crank area
Flintstone opening
Body panels aft of operator
Output impeller frame mount redraw

Resize head areas of PTUs
casings and head inlets

6.5025
3.64
5.262
8.511

17.798
4.5° 80x
1.798
3.00
24.00
17.00
1.798
10.00
24.00
.936* .553
.386*
7.00

Sherline
Backpacker's self propelled

~~1.798~~

6.202

3.101

↔ 9.241 12.00 ↕

.358 ↓ 0.358 —

8.142

4.014

10.00" ↕

$\frac{16.238}{2}$

~~61.494627°~~ 68.15875°

40.004796°

41.632792°

← 6 → $\frac{32.238}{2}$ ——

↑
4
↓

10.508
3.114 ↓
8.114 →

66,965.6 lbs. Tractor Cab

275/80R24.5 42.1 in overall dia. 12.2 min. dual Spacing

Max 75 mph 10.8 in wide

42.1 • π = 132.26105 in / 12 in/ft 5675 lbs dual max

trailer

75 mph × 44 ft/sec / 30 mi/hr = 110 ft/sec 41.2

steer

110 / 11.02170 = 9.98026 Revs/second 41.4

117,468.8234 lbs

53.9 • 71.44785 3in / 12^3 3in/3ft 2.37374

lb/3ft

−2n/2n + n = 15.60856 20.51302 117,468.8234 × = 450

×

234,937.6467 lbs / 2 = 117,468.8

21.05

66,965.6 lbs / 2

.85036

1.19366

1.7188

1.21536

1.2154

117,468.8234 lbs × 1.700

5 impellers × 750 lbs each

= 2.2198 12 in/ft

4.42954 ft radius

primary impellers (5x)

1.1188 dia

7.27786r

2.37374

71.44785 3in

120.51302

output impeller gallery

971.9791

2.2045 ^{2}in +
~~3.23308~~ ^{2}in × 3 = 9.69924 ^{2}in

~~11.90374 ^{2}in~~
8.818

~~22~~ 971.9

11.90374 ^{2}in · 8 ~~= 969.99544 in~~

13.77763 in 8 Rev PTU 81.64906 in

2.5515 in = 10.20613 in gear casing depth.
4 PTU

in × depth in × 30,000 lbs/^{2}in = 117,468.8234 lbs

$(6.28125^2 \cdot \pi - 6^2 \pi) \cdot .21875 = 2.373743$ in

$\frac{81.64906}{12\ \text{PTUs}}$ = 6.80409 in
16

5.10307"

3 primary impellers
16 secondary PTUs
w/radius of
~~26.6376~~ in
~~133.18772~~
44.39559 in

$\frac{117{,}468.8234\text{ lb} \times .85036\text{ in}}{44.39559\text{ in}}$ = 750.00534 lb

/3 primary impellers

$(44.6\,\frac{1429}{}^2\pi - 44.39559^2\pi) \cdot .2812 = 17.19698$ 3in

$\frac{44.39559 \times 2 \times \pi \times 30000 \times n}{12^3\ \text{3in/3ft}} = 750.00534$ lbs. × 3 primary impellers

$n = .15487$

$\frac{2250.01602\ \text{lbs.}}{53.9\frac{lb}{ft^3} \frac{17.19698\ \text{3in}}{12^3\ \text{3in/3ft}}} - \frac{2n}{2n} + n = 12.02407$

17.19698 3in · 12.02407 · 2 · 8 Rev's. = 3,308.44306 3in.

primary impeller volume · fold exponent · double hydraulic volume · @ primary impeller

$\frac{\frac{3308.44306\ \text{3in}}{8.818\ \text{2in}}/2\ \text{Rev per crank c/k}}{8\ \text{revs}}$ = 23.4495 inches primary [illegible] casing depth.

$\frac{2250.0160}{46}$ [illegible] .5036 [illegible] rank = chainring radius of 47.83309 inches

23.4495 × 5.7797 = 22.58831 = 5.7797

47.83309 / 52π · .5/π

Make six → twelve sets of four primary PTUs.

$\frac{22.58851}{6}$ inches primary gear casing depth for 1 PTUs @ full balance.

= ~~7.5295 inches~~ × 6 PTU assemblies. × 2, one for each primary impeller

1 × 8

× 2

= 12

Slip equation(s)

$\frac{23.4495}{3}$ × 5.7797 sprocket ratio

Stacked 3 high

$\frac{24}{2}$ PTUs for each primary impeller (2×)

1-312-878-9013

#3128789013

one PTU

Conclusion

If this publication does not make things so I can file a patent application too, as well as anyone else, I will be surprised. I hope this publication provides you, the reader, with some useful information and perhaps you have learned something you didn't know or would have known had you been able to find the information somewhere else. I hope, with understanding the elementary simplicity of this first and original idea and discovery, that even a child will be able to write the books left to be written, and that the understanding of base two mathematics will lead to some interesting revolutions in software and computer technology. Thanks for purchasing my publication on Human Powered Transportation Means. I hope it will continue to prove useful in any endeavors of any kind you hope to attain.

Richard Chastain

Printed in the United States
By Bookmasters